RESPONSIVE TEACHING

This essential guide helps teachers refine their approach to fundamental challenges in the classroom. Based on research from cognitive science and formative assessment, it ensures teachers can offer all students the support and challenge they need - and can do so sustainably.

Written by an experienced teacher and teacher educator, the book balances evidence-informed principles and practical suggestions. It contains:

- A detailed exploration of six core problems that all teachers face in planning lessons, assessing learning and responding to students.
- Effective practical strategies to address each of these problems across a range of subjects.
- Useful examples of each strategy in practice and accounts from teachers already using these approaches.
- Checklists to apply each principle successfully and advice tailored to teachers with specific responsibilities.

This innovative book is a valuable resource for new and experienced teachers alike who wish to become more responsive. It offers the evidence, practical strategies and supportive advice needed to make sustainable, worthwhile changes.

Harry Fletcher-Wood is Associate Dean at the Institute for Teaching. He has spent a decade in schools as a classroom teacher, head of professional development and as a researcher. He blogs regularly at improvingteaching.co.uk and tweets as @HFletcherWood.

RESPONSIVE TEACHING

Cognitive Science and Formative Assessment in Practice

Harry Fletcher-Wood

LONDON AND NEW YORK

First published 2018
by Routledge
2 Park Square, Milton Park, Abingdon, Oxon OX14 4RN

and by Routledge
711 Third Avenue, New York, NY 10017

Routledge is an imprint of the Taylor & Francis Group, an informa business

© 2018 Harry Fletcher-Wood

The right of Harry Fletcher-Wood to be identified as author of this work has been asserted by him in accordance with sections 77 and 78 of the Copyright, Designs and Patents Act 1988.

All rights reserved. No part of this book may be reprinted or reproduced or utilised in any form or by any electronic, mechanical, or other means, now known or hereafter invented, including photocopying and recording, or in any information storage or retrieval system, without permission in writing from the publishers.

Trademark notice: Product or corporate names may be trademarks or registered trademarks, and are used only for identification and explanation without intent to infringe.

British Library Cataloguing-in-Publication Data
A catalogue record for this book is available from the British Library

Library of Congress Cataloging-in-Publication Data
Names: Fletcher-Wood, Harry, author.
Title: Responsive teaching : cognitive science and formative assessment in practice / Harry Fletcher-Wood.
Description: Abingdon, Oxon ; New York, NY : Routledge, [2018] | Includes bibliographical references and index.
Identifiers: LCCN 2017060136 (print) | LCCN 2018014627 (ebook) | ISBN 9781315099699 (ebook) | ISBN 9781138296879 (hbk) | ISBN 9781138296893 (pbk) | ISBN 9781315099699 (ebk)
Subjects: LCSH: Effective teaching. | Cognitive learning. | Lesson planning. | Educational tests and measurements.
Classification: LCC LB1025.3 (ebook) | LCC LB1025.3 .F59 2018 (print) | DDC 371.102–dc23
LC record available at https://lccn.loc.gov/2017060136

ISBN: 978-1-138-29687-9 (hbk)
ISBN: 978-1-138-29689-3 (pbk)
ISBN: 978-1-315-09969-9 (ebk)

Typeset in Interstate
by Apex CoVantage, LLC

In memory of Claire Hartnett

CONTENTS

FIGURES

ACKNOWLEDGEMENTS

I have benefited enormously from the support, suggestions and challenges of friends, colleagues and early readers.

Contributions by Damian Benney, Jason Chahal, Robin Conway, Rowan Pearson, Michael Pershan, Lizzie Strang and Warren Valentine brought critical ideas to life: thank you.

Marcus Bennison, Adam Boxer, Jon Brunskill, Emma McCrea and James Theobald allowed me to share examples of their work: thank you.

My understanding of key ideas was transformed, as was the book, by conversations with Daisy Christodoulou (success criteria), Peps McCrea (focusing on problems), Michael Pershan (feedback) and Gaurav Singh (prototypes and toolkits): thank you.

Valuable ideas, suggestions and encouragement came from Steve Adcock, David Barclay, Jen Barker, Christian Bokhove, Adam Boxer, Melissa Christey, Pie Corbett, Zeba Clarke, Lucy Crehan, Sarah Donarski, Toby French, Josh Goodrich, Nick Hassey, Clare Hill, Sara Hjelm, Niki Kaiser, Ollie Lovell, Ben Moss, Lisa Pettifer, Arthur Reeves, Emma Saadatzadeh, Roo Stenning, Phil Stock, Ben White, Henry Wiggins, Dylan Wiliam and Routledge's anonymous reviewers: thank you.

David Didau and Joe Kirby's scepticism did not convince me, but it pushed my thinking.

Isabel Buckles, Emily Hammond, Will Hopkinson, Yusuf Mohammed and Paul Sims tested many of these ideas and showed me what guidance might be helpful: thank you.

These ideas were refined through presentation to, and questions from, trainees and colleagues at Ark Teacher Training and delegates at the 2017 Historical Association Conference.

The Institute for Teaching has provided the ideal blend of challenge and support: thank you especially to Marie Hamer, Matt Hood and Peps McCrea.

I'm grateful to Clare Ashworth at Routledge and to Annamarie Kino, whose suggestion this book was.

Most importantly, thank you Loren.

FOREWORD

The realisation that students do not learn what they are taught has probably been around since humans first started trying to teach each other anything. This, combined with the idea that good teaching begins from where the learner actually is, rather than where the teacher would like her or him to be, means that good teaching has always included some way of checking for the sense that learners are making of their learning experiences. In her seminal works on the practice of teaching (see, for example, Hunter, 1982), Madeline Hunter stressed the importance of frequent "checks for understanding" in teaching, and this is something that most teachers do intuitively.

Independently of this, around fifty years ago, educational psychologists such as Benjamin Bloom began to look at ways that the periodic assessments of student achievement via tests and examinations - which again have been part of the landscape of education for centuries - might be used to improve students' classroom experiences (Bloom, 1969). A couple of years earlier, Michael Scriven had proposed that when a new curriculum or some other innovation was evaluated, this might be done with a view to improving the curriculum while it was being developed or to determine whether the new curriculum, once developed, was worth implementing more widely. Scriven also suggested that the words 'formative' and 'summative' should be used to describe these two different roles that evaluation might play (Scriven, 1967). Much to Scriven's irritation, Bloom suggested that the terms 'formative' and 'summative' could also be applied to the evaluation of individual students.

It seems that Bloom's proposals gained little traction in schools on either side of the Atlantic, but, particularly in the UK, many universities began to incorporate 'formative assessments' into their courses, although in many cases, such assessments amounted to little more than 'any assessment before the big one.' The predominant usage of the term 'formative assessment' was to describe formal assessments that in some way mimicked the assessments students would take at the end of a course of study, usually intended to do little more than inform students how they were doing and, specifically, what sort of score they would get.

However, while the term 'formative assessment' tended to be used to describe formal, 'set-piece' assessments at the school level, the idea that something called 'assessment' might be useful in deciding what to do next with students was creating a great deal of interest. In the 1960s, Bertram Banks designed the Kent Mathematics Project (KMP) on the principle that students would be assigned individual work and complete a formal test on that work, with the difficulty of the next batch of assignments based on how well a student did on the test (Banks, 1991). Where the score was below a pre-defined level of 'mastery', the next batch of assignments would be pitched at a less demanding level, or provide consolidation of the material on which the student had struggled. Where the student did reasonably well, the next batch would be slightly more challenging, and where the student did very well, the teacher might set substantially more difficult work. Largely because of the difficulty of licensing the KMP materials, in the early 1970s, teachers in Inner London Education developed a similar scheme - the Secondary Mathematics Individualised Learning Experiment (SMILE), which was used extensively in England throughout the 1970s and 1980s (Gibbons, 1975).

In parallel, teachers of children with special educational needs had been realising the power of assessment to help them make better decisions about the learning needs of their students. For many years, the assessment and diagnosis of students with special educational needs had been dominated by the use of intelligence tests, but many teachers and researchers had become increasingly dissatisfied with the evidence such assessments yielded about students' learning needs, and looked for alternative procedures that might actually inform, rather than just measure, learning:

> Such procedures are designed to provide information on the processes and strategies which an individual uses to solve problems and deal with specific situations. These processes may be no different in kind from those used by other people, whether normal or handicapped, but may nevertheless call for rather specific methods of remediation if the individual is to be helped in the direction of normality or at least to the next stage of his own development.
>
> (Mittler, 1973, p. 14)

In the 1980s, three reviews of research examined the impact of classroom assessment practices on students. The first, by Lynn and Douglas Fuchs, reviewed research on students with special educational needs and, specifically, focused on studies that examined the impact of regular assessment on student progress (Fuchs and Fuchs, 1986). The other two looked more broadly at the impact of classroom assessment practices on student achievement and motivation (Crooks, 1988; Natriello, 1987). All three studies confirmed that assessment could improve student learning but often did not, for a variety of reasons.

In 1997, Paul Black and I reviewed studies undertaken in the subsequent decade and confirmed the findings of the earlier reviews: using assessment to guide classroom activities could produce substantial improvements in learning (Black and Wiliam, 1998a). We also produced a non-technical summary of our findings, intended for practitioners and policy makers, titled "Inside the black box," which was disseminated widely (Black and Wiliam, 1998b).

Paul Black and I had used the term 'formative assessment' to describe assessment that was actually used to modify the classroom activities in which students were engaged, primarily because we wanted to rehabilitate the word assessment and point out that the most important assessments happened during teaching, not after it. Others preferred the term 'Assessment for Learning', not least because formative assessment was still widely associated with the use of formal assessments.

This distinction might not have mattered much, except in the late 1990s, the Labour government embarked on a series of ambitious reforms of the teaching of 11- to 14-year-olds in secondary schools in England, titled the "Key Stage 3 Strategy." Building on the National Strategies for Literacy Numeracy, which had been introduced into primary schools shortly after the government came into power in 1997, the package of reforms for secondary schools included specific programmes for the teaching of English and mathematics - but for the other 'foundation' subjects in the National Curriculum, the Strategy included three main strands: structured lessons, thinking skills, and Assessment for Learning (Stobart and Stoll, 2005).

However, while the Key Stage 3 Strategy claimed to be based on research, the approach to Assessment for Learning adopted bore little relation to the research discussed above. Rather, Assessment for Learning was defined primarily as a process of monitoring student progress, using increasingly sophisticated spreadsheets, with national expectations of the numbers of levels of progress a student should make.

The result is a deeply frustrating situation. Research showing the profound impact that minute-to-minute and day-to-day assessment can have on student achievement continues to accumulate. And yet, the terms that we have to describe these processes - formative assessment and Assessment for Learning - are inadequate to describe them, not least because people have such fixed ideas about the meaning of these terms.

In my own work, I have stuck to the term 'formative assessment' and tried to get people to understand what I mean by the term. However, after over twenty years, I have to concede that this may not be the best approach, and a new terminology may be needed. I just don't know.

What is clear to me, after a quarter of a century of supporting teachers in developing their use of assessment to support their students' learning, is that working out exactly how to help teachers respond to the learning needs of their students involves moving away from the labels and getting into the detail of classroom practice, and that is why I was delighted to be invited to write the foreword to this extremely important book.

Through the use of vignettes, stories and teachers' own accounts, Harry Fletcher-Wood takes us on a journey from the tokenistic adoption of Assessment for Learning that was so common in the Key Stage 3 Strategy, to a deep understanding of how we can make our teaching responsive to our students' needs. I know of no other book that comes close to this in helping teachers really understand, at the deepest level, what makes teaching responsive to students and how every teacher can use these ideas to improve their practice.

Dylan Wiliam

Introduction

Getting responsive teaching wrong

I taught poorly for a long time, because three things confused me: assessment seemed to hinder learning, skills seemed more important than knowledge and Assessment for Learning seemed to be just a collection of techniques. My confusions are now only a historical curiosity, but what I discovered about learning and assessment as I overcame them, slowly and reluctantly, may be of more lasting value: this is what *Responsive Teaching* shares. To introduce what I discovered, I should begin by describing why I was confused.

Confusion 1: assessment seemed to hinder learning

As a new teacher, I could not see how assessment supported learning. Our assessments were regular, graded tests. We asked Year 9 to evaluate evidence about conditions in Victorian factories: this meant two lessons helping students answer eight complicated questions, then an evening writing a comment and a grade for each question. Few students received higher grades, perhaps because we taught them nothing about using evidence and nothing about Victorian factories before the assessment. Assessment did not seem focused on what students were learning.

Marking those assessments meant applying a measurement system which was superficially precise, but practically incoherent. We applied levels, specified by the National Curriculum, to different historical skills. In explaining why events happen, at:

- **Level 5:** students begin to make links between causes.
- **Level 6:** students make links between causes.

Many schools split levels into sub-levels:

- **Level 5c-:** students begin to begin to make links between causes.
- **Level 5a:** students always begin to make links between causes.

My school split sub-levels into sub-sub-levels for greater precision: 5a+, 5a, 5a- and so on. Presumably, therefore:

- **Level 5c-:** students begin to begin to begin to make links.

This incredibly precise system totalled eighty-one steps (nine levels, each split into nine sub-sub-levels) and yet I struggled to separate a 'Level 5' from a 'Level 6' answer. The marks we gave students seemed detached from their learning.

I was under pressure to improve students' reported levels: their levels increased accordingly, but their knowledge and skill did not. Levels were designed to assess students' work at the end of three or four years' learning: instead, we awarded them based on answers to individual questions. One good answer about conditions in Victorian factories justified giving a student a Level 6, even if they knew nothing else about the curriculum. Two students at 'Level 6' might know completely different things (Department for Education, 2015): one might be able to use evidence about Victorian factories well; another might have written a good essay explaining the significance of Magna Carta, but know nothing about Victorian factories. Our assessments seemed to reflect the progress we wanted students to make, not what they actually knew.

Analysing assessment results was even further detached from reality. I entered a sublevel in my markbook, then copied it into the school's data management system; the head of department wrote a report analysing these imprecise results in minute detail; the headteacher returned it with questions: why had two students not reached the levels our data tracking system expected? One was in hospital during the previous test; the other had not attended school since September. It was surprising that the other twenty-three students made expected progress, given that only 9% did so nationally (Treadaway, 2015). "Vastly over burdensome systems" to analyse results absorbed time (Department for Education, 2015, p. 31) and invited dishonesty. Assessment was detached from what I taught, it was detached from what students learned and our analysis was detached from reality: I saw it as a hindrance.

Our problems stemmed from misusing assessments designed for summative purposes. Summarising learning is important: students applying for jobs or further study need to show what they have learned; their qualifications must have a shared meaning, agreed by schools, employers and universities. In selecting Ellen over Edward, employers need confidence that her '6' in GCSE Maths reflects greater knowledge than his '4' – not luck, lenient marking or additional help. Employers can be confident only if Ellen and Edward sit exams which include questions testing the whole syllabus, in the same time, with the same help, marked to the same standard (Christodoulou, 2017). These are stringent conditions, but we need meet them only when students have reached the end of a course: what Ellen and Edward knew halfway to completing their GCSE is of limited interest to employers and universities. Yet three factors led us to misapply summative assessments to everyday teaching.[1]

Factor 1: we want to know how students are doing

We misapplied assessments designed for summative purposes to everyday teaching because practice exams seem a logical way to check if students are on track. However, an exam testing students' knowledge of the entire maths syllabus in only ninety minutes is a "blunt and insensitive instrument" with which to measure progress over a term (Christodoulou, 2017, p. 122). Students may master fractions over a term but do no better in a practice exam, because their new knowledge helps on only a few questions. If we collect exam questions about fractions, our test no longer demands knowledge of the whole syllabus: it shows whether students can answer those questions, but not how well they would do in an exam. Without standardised questions, conditions and marking, grades from practice summative assessments tell us little about how students will do in their final exams. Yet we used practice exams regularly and treated the results as valid. Assessments seemed detached from learning and students showed little improvement, because I was applying questions and mark schemes designed for the end of the course six times a year.

Factor 2: we want to teach students the right things

We misapplied assessments designed for summative purposes by using them to guide what we taught, hoping this would help students improve. Teaching what will be on the test seems rational, if reductive, but it undermines students' learning. Learning requires "lots of specific knowledge and lots of deliberate practice at using that knowledge" (Christodoulou, 2017, p. 42). The tasks we want students to master eventually often differ from those which build knowledge and offer practice: in maths, we want students to be able to solve problems, but practising solving problems is an inefficient way to learn mathematical concepts and become a better problem solver (Sweller, van Merriënboer and Paas, 1998). Writing an essay about *An Inspector Calls* demands students integrate quotations and insights about character, theme and social context into a coherent argument: doing so robs students of practice in each individual component (Quigley, 2017). Summative assessments may influence long-term planning, but we used them to designs lessons which allowed students to "learn or demonstrate the requirements for specific levels" (Department for Education, 2015, p. 13). Christodoulou (2017) compares this with marathon training: training for marathons by practising marathons would be exhausting, inefficient and would neglect the strength and conditioning runners need. Exams are proxies for achievement: teaching students to meet their demands – 'knowing how to answer a four-mark question', for example – focuses on superficial features, not genuine understanding of the subject; this constrains students and lowers our expectations (Massey, 2016). My attempts to plan student learning using National Curriculum levels foundered because they drew me away from the practice and knowledge students needed.

Factor 3: we want to identify gaps in students' learning

We misapplied assessments designed for summative purposes because we hoped they would identify gaps in student learning: we should have used assessments designed for this formative purpose. Exam questions designed for summative purposes draw on knowledge of the whole syllabus, so they offer limited formative information:

> Some pupils will fail to answer a question about the area of a rectangle because they don't understand how to calculate the area of a rectangle. Others will struggle because their knowledge of basic number facts and times tables is weak. Even if pupils show their working, it is hard to infer reliably exactly what the source of their struggles are.
>
> (Christodoulou, 2017, p. 115)

Similarly, a student may write poorly about Brontë's presentation of Jane Eyre because they struggle with specific vocabulary, structuring paragraphs or knowledge of contemporary expectations of women. Exam mark schemes are fit for their purpose, applying a summative grade to an exam; this limits their usefulness in identifying how students can improve

(Massey, 2016). Students, parents and teachers focused on National Curriculum levels, not how students could improve (Department for Education, 2015); a student's complaint, "I want to be a Level 4," exemplifies the unproductive distraction this caused. Assessments designed for formative purposes would have been more useful: they elicit evidence of students' understanding, which allows teachers and students to make better decisions about next steps (Black and Wiliam, 2009, cited in Wiliam, 2016, p. 106). They need not meet the strict conditions summative purposes require: standardised questions, support and marking. The stakes are lower: formative assessments lead to a consequence, a response by a teacher or student; we want to make valid inferences about what students have learned, but if our inferences are wrong, students' responses will make this clear (Christodoulou, 2017). Rather than identifying what students had learned using formative assessment, however, focusing on summative assessments left us "simply tracking pupils' progress towards target levels" (Department for Education, 2015, p. 13). Analysing the results of tests designed for summative purposes offered limited information about the gaps in students' learning and distracted us from assessments for formative purposes.

Assessments for summative purposes dominated my teaching, my leaders' expectations and my students' thoughts: this hindered learning. We hoped to track student success, but generated inaccurate information; we hoped to plan learning, but obsessed over superficial exam features; we hoped to identify gaps, but learned little of use. Practice exams can increase students' confidence about exam structure and timing, but this is valuable only just before the final exam; at any other point, summative assessments are a distraction. We should confine assessments designed to summarise what students have learned to their purpose; we should use assessments designed for formative purposes to track students' progress and identify gaps. We should plan what students are to learn based on what success looks like in our subject, not the exam mark scheme. If we plan for success in our subject and get formative assessment right, students will learn (and so be prepared for summative assessments); this requires escaping the confines of summative assessments and the problems they can cause: warped priorities, wasted time and inaccurate diagnoses.

Responsive teaching uses assessments for formative purposes to identify what students have learned: it avoids the distraction and distortion assessments for summative purposes can create. The principles of responsive teaching apply whatever summative assessment students will take; they will continue to apply after the next curriculum change. Teachers may wish to use summative assessments more frequently, or may have no choice, but limiting their influence on teaching helps us focus on what students know and how we can help them improve. Recognising the limits of summative assessments only overcame my first confusion, however; our assessments also confused me about how students learn.

Confusion 2: skills seemed more important than knowledge

I questioned our assessment system, but I accepted one underlying premise uncritically: a focus on skills over knowledge. The National Curriculum specified how to improve at six historical skills; it did not specify the knowledge needed. One historical skill was explaining why events happen: students could reach Level 5 explaining the causes of the English Civil

War or the First World War. They could reach Level 5 even if they knew nothing about either. Teaching a group of Year 9 students with little interest in history, I reasoned that they could also reach Level 5 by explaining the causes of the 2011 London riots. The focus was "getting pupils across the next threshold," rather than "ensuring they were secure in the knowledge and understanding defined in the programmes of study" (Department for Education, 2015, p. 5). I could not teach students causation without including historical events, but I believed procedural knowledge - skills students could use elsewhere, like arguing persuasively - was more valuable than knowledge of history.

Beyond historical skills, I wanted my students to gain transferable skills which would help them lead better lives. I worked at two schools successively which adopted a programme of thinking and learning skills across subjects, such as 'reasoning', 'distilling' ideas and 'making links' between them. I embraced this wholeheartedly. I believed that skills such as creativity, collaboration and critical thinking would prepare students for the real world; rather than teaching history, I dedicated lesson time to encouraging students to 'reason' and 'manage distractions'. First, I wanted my students to be happy, engaged and motivated; if this happened, they were succeeding. Second, I wanted them to learn independently and gain skills which would be useful no matter what they ended up doing. My third priority was teaching history. It took me a long time to realise that neglecting history undermined the other two goals.

Researchers spent decades trying to isolate the general thinking skills I sought to teach; their findings suggested that skills depended on knowledge. Originally, researchers believed expertise meant thinking faster or better: being good at decision making, inventive thinking and problem solving (Perkins and Salomon, 1989). They believed that while experts needed some knowledge of the context - a chess master needed to know the rules of chess - they could improve in chess by studying military strategy or other board games. A breakthrough study by Simon and Chase (1973) questioned this. It built on the work of Adriaan de Groot, who found that chess experts can remember game situations at a glance, but novices can remember the location of only a handful of pieces. Simon and Chase found that if the pieces are placed at random, however, masters can remember little more than novices. They showed that chess masters learn - memorise - patterns of pieces. They take in game situations at a glance because they recognise them; they win games not because they think faster, but because their knowledge allows them to recognise patterns "quickly and unconsciously, and the plausible move comes almost automatically" (Simon and Chase, 1973, p. 403). Building on this work, belief in general thinking skills was undermined by three bodies of evidence (Perkins and Salomon, 1989):

1. Experts know a huge amount about their field: like chess masters, they know patterns, recognise when they apply and reason accordingly.
2. General problem solving strategies are inefficient and often fruitless: knowing you need to break down a problem into smaller steps is unhelpful unless you know possible steps and how to complete them.
3. Skills transfer poorly, if at all, between contexts: good chess players are not particularly good at other strategic games, such as Go.

I believed students could learn skills and apply them to any knowledge. The reverse was true: students could show skill only where they had knowledge. Critical thinking, for example, depends on knowing the field and what success looks like within it (Bailin et al., 1999). Solving maths problems means identifying the operations to perform and doing so correctly; we may call this 'problem solving', we may want students to do it 'thoughtfully', but teaching problem solving (or thoughtfulness) does not help unless students recognise the problem and know the solution steps. We can teach students to think deeply, critically and creatively, but only about what they know. I embraced the approach the assessment system promoted and missed the importance of what students know: failing to teach my students historical knowledge doomed my attempts to teach them to think skilfully.

The importance of acquiring and organising knowledge proved a useful way to think about teaching. Learning means committing new elements to memory: chess masters learn positions, historians learn events, doctors learn symptoms and cases. They do not just accumulate facts, however: experts' knowledge is organised and usable. This allows experts to focus on salient points, understand and approach problems effectively, think more efficiently and respond automatically (Chi, Glaser and Rees, 1982; Klein, 1998; Larkin et al., 1980). Doctors' knowledge allows them to diagnose rapidly and accurately, for example (Schmidt and Rikers, 2007). To help students become experts, we need to teach students what experts know, rather than expecting them to act as experts without the necessary knowledge. We want students to analyse, debate and solve problems, but this relies on their knowledge and how students see the connections between words, ideas and concepts. Comparing Othello with Hamlet relies on students knowing Hamlet's words and actions well enough that they can discuss his character without having to remind themselves of who he was and what he did. Learning means not just committing new elements to memory, but organising and connecting what we know, and automating smaller tasks to allow us to concentrate on bigger ones. This is how students gain expertise in a subject. Alfred North Whitehead argued that this is how civilisation advances, through "extending the number of important operations which we can perform without thinking about them" (1911, p. 34). The importance of knowledge has three consequences for responsive teaching.

Consequence 1: what students already know matters

What students know dictates what they can learn. Students with low reading ability but good knowledge of baseball will understand a text about baseball as well as students with high reading ability and a good knowledge of baseball, and better than students with high reading ability and a low knowledge of baseball (Recht and Leslie, 1988). Students can make sense of the Reformation only with knowledge of Tudor monarchs, Christian worship and sixteenth century European politics. Their misconceptions - incorrect beliefs - influence what they learn, too; students who believe germs are smaller than atoms may struggle to understand both. If we know what students know, we can plan what to remind them of, what to build upon, what to seek to correct; we can pitch our lessons appropriately, identify who needs help and assess who has learned what. David Ausubel described students' existing knowledge as the "most important single factor influencing learning . . . Ascertain this and teach him accordingly" (Ausubel, 1968, in Wiliam, 2016, p. 101). Responsive teaching recognises that we must plan based on what students know, since it dictates what they can learn.

Consequence 2: students need knowledge and skill in the subject

Developing expertise means gaining knowledge and skill in a subject. This requires identifying the building blocks of knowledge and skill, sequencing them carefully and ensuring students gain and retain them. This process, known as deliberate practice, is *the* route to improving performance, taken by experts in fields as varied as chess, memorisation and musical performance (Ericsson and Pool, 2016). Formative assessment has been criticised for focusing on general skills and techniques, overlooking the substance students are thinking about and the value of domain-specific knowledge (Bennett, 2011; Coffey et al., 2011). Responsive teaching requires planning and sequencing what students are to learn: responding to what students have learned is worthwhile only if the curriculum is rigorous, challenging and carefully structured.

Responsive teaching recognises that skill and success rely on what students know. This entails:

- **Identifying what students already know.**
- **Planning and sequencing learning based on the knowledge we hope students will gain.**

Planning learning is essential, but it does not guarantee learning. The third consequence of the importance of knowledge is the need to check what students have learned. Assessment for Learning should have helped, but instead it was my third area of confusion.

Confusion 3: Assessment for Learning was a bunch of techniques

The importance of knowledge means we have to check whether students have learned and remembered what we taught them, both for its own sake and because this will dictate whether they can learn the next topic. Planning and sequencing teaching matter, but they are insufficient, because what students learn is unpredictable. Graham Nuthall tracked student learning meticulously, recording what individuals said to their peers and themselves, reading everything they wrote and interviewing them about the origin of their ideas; he concluded: "No matter how well you describe something, how well you illustrate and explain it, students invent some new way to misunderstand what you have said" (2007, p. 24). Variation in prior knowledge and in what students attended to meant that "what students learn during a unit is largely unique. Even though they are in the same class and apparently engaged in the same activities, what one student learns is not the same as what other students learn" (p. 100). Good curriculum planning is insufficient because teachers cannot control "students' environments to the extent necessary for unintended conceptions not to arise" (Wiliam, 2011, pp. 74-75): what students watch, what they hear at home or the explanations they invent for their experiences. Even if we thought we had created a perfect curriculum and predicted students' environments perfectly, we would still need assessment to check we were correct. Assessment for Learning should have promoted this focus on what students were learning, but my use of it did not achieve this.

I believed I was doing Assessment for Learning well: I was not alone in approaching it wrongly. I embraced techniques: using lollipop sticks to select students to speak, having students write paragraphs on mini-whiteboards, experimenting with ways to engage students with the learning objectives. My enthusiasm for techniques was not matched by my understanding of the underlying principles, however. I shared objectives, but they were constructed hurriedly and uncritically. Students used mini-whiteboards, but I could not read thirty paragraphs at once: I was eliciting writing, not evidence of students' learning. My marking was detailed but infrequent: I had little sense of what students had learned each lesson. I was not alone in this; Assessment for Learning often seemed to prioritise techniques isolated from students' learning (Coffey et al., 2011). Many teachers, such as Joe Kirby, came to see it as a collection of gimmicks, not a group of principles:

> Lolly pop sticks, coloured party cups, red-amber-green traffic lights five times a lesson, thumbs-up-or-down, starred self-confidence post-its, scribbled emoticons for end-of-lesson feelings, strange and unhelpful acronyms like WALT & WILF all became a kind of reductio ad absurdum. Many senior leadership teams then enforced the letter of the AfL law rather than the spirit of it: school-mandated lesson plans, observation rubrics and progress checks 3 times a lesson and endless mini-plenaries; objective sharing in rigid but often counterproductive formats required across all subjects like 'by the end of the lesson, students will be able to . . . '; peer assessment on levels that often resulted in comments like: '5a because he tried hard and wrote neat'; marking in green pen rather than red to avoid damaging students' self-esteem; and posters with tiny, illegible, incomprehensible but displayed level descriptors. Prescriptive but flashy AfL techniques like waving around mini-whiteboards became the OFSTED-enforced orthodoxy, and inspectors became obsessed that pupils could say what level they were on.
>
> (Kirby, 2014)

A gap had emerged between the 'letter' of teacher's actions and the 'spirit' underpinning Assessment for Learning (Marshall and Drummond, 2006). As a result, much Assessment for Learning practice in schools applied techniques which achieved little, detached from the principles Paul Black and Dylan Wiliam sought to promote.

This was a shame, because formative assessment, embodied as Assessment for Learning, promised much. Students seem likely to learn more when evidence of their achievement

> is elicited, interpreted, and used by teachers, learners, or their peers, to make decisions about the next steps in instruction that are likely to be better, or better founded, than the decisions they would have taken in the absence of the evidence that was elicited.
>
> (Black and Wiliam, 2009, in Wiliam, 2016, p. 106)

The most recent round of PISA international tests found that

> the biggest impact on student achievement in science was the affluence of the parents. The second was the ability of teachers to adapt instruction to meet student needs. . . . There is literally nothing else that can increase student achievement by so much, for so little cost.
>
> (Wiliam, 2018)

Critics have questioned the impact of formative assessment on student learning (Bennett, 2011), yet recent reviews have demonstrated substantial impact (Kingston and Nash, 2011) and specific approaches to formative assessment have strong evidence bases (each chapter discusses the evidence for the approach suggested). Assessment for Learning had promise.

Despite this promise however, student learning did not improve. Rob Coe has argued that it is now rare

> to meet any teacher in any school in England who would not claim to be doing Assessment for Learning. And yet, the evidence presented above suggests that during the fifteen years of this intensive intervention to promote AfL, despite its near universal adoption and strong research evidence of substantial impact on attainment, there has been no (or at best limited) effect on learning outcomes nationally.
>
> (2013, p. x)

The Assessment Commission concluded that, almost twenty years after the introduction of Assessment for Learning, "formative classroom assessment was not always being used as an integral part of effective teaching" (Department for Education, 2015, p. 13). There are competing explanations as to why promising results from pilot studies were not replicated nationally. The distortions caused by the summative assessment system and the neglect of subject knowledge have been discussed. Sue Swaffield (2009) criticised implementation through government directive, leaving teachers at the bottom of a chain of command. Randy Bennett (2011) highlighted unclear definitions of formative assessment and insufficient teacher knowledge of assessment, content and pedagogical techniques. Changes in classroom assessment need to align with other influences on education, such as exams and teacher training; this was not the case (Cambridge Assessment, 2017; Bennett, 2011). More generally, projects which are successful locally often lose their power without the support and focus of their initiators (for a vivid example, see Yeager and Walton, 2011, p. 288). Nationally, plausible principles failed to achieve their promise; locally, I thought I was doing Assessment for Learning, but this had little impact on my students' learning.

Responsive teaching focuses on the principles of formative assessment, not the techniques; it benefits from the experience accumulated over twenty years of Assessment for Learning. Overcoming my three confusions introduced me to the limits of summative assessment, the merits of formative assessment and the guidance of cognitive science; this forms the foundation of responsive teaching.

What is responsive teaching?

Responsive teaching blends planning and teaching, based on an understanding of how students learn from cognitive science, with formative assessment to identify what students have learned and adapt accordingly. Dylan Wiliam has suggested that 'responsive teaching' might have been a better term for Assessment for Learning. This seems apt, focusing on what students are thinking and how we respond (Coffey et al., 2011). It emphasises the interactive nature of classroom teaching, the demands on the teacher and what success looks like. A new term encourages a fresh view of formative assessment, free of the influence of summative assessment and the superficial techniques which discredited Assessment for Learning. Responsive teaching benefits from the research evidence and practical wisdom which has accumulated around cognitive science and formative assessment, while seeking to overcome some of the confusions from which I suffered. Responsive teaching:

- Addresses Confusion 1 by distinguishing between formative and summative assessment and focusing on what students have learned and how they can learn more.
- Addresses Confusion 2 by focusing on how students learn: acquiring and organising knowledge and skill in a subject.
- Addresses Confusion 3 by focusing on the principles of formative assessment and using the practical wisdom gained from the introduction of Assessment for Learning.

The book shows these ideas in practice by focusing on six endemic problems. Endemic problems are predictable, inevitable and intrinsic challenges in teaching (Lemov, 2015), in contrast to exotic one-offs, like the chaos caused by a wasp's visit to my Year 11 lesson. The problems are:

1. How can we plan a unit, when we want students to learn so much, and have so little time?
2. How can we plan a lesson, when we want students to learn so much, and have so little time?
3. How can we show students what success looks like?
4. How can we tell what students learned in the lesson?
5. How can we tell what students are thinking?
6. How can we help every student improve?

No matter how good our leaders and how diligent our students, these problems will always be challenging. Conversely, teachers can refine their approach to each problem, no matter what challenges their school faces and what leaders and politicians prioritise. The problems may be answered in sequence: we need clear plans if we are to show students what we expect; we need good assessments if we are to help students improve. This is why the first two questions relate to planning: good planning is a prerequisite to worthwhile response (this also overcomes a major criticism of formative assessment; see Bennett, 2011; Coffey et al., 2011). This also offers a sequence for improving as a teacher, a critical aspect of deliberate practice and effective coaching (Ericsson and Pool, 2016; Deans for Impact, 2016).

How does this book help teachers adopt responsive teaching?

Adopting responsive teaching means grasping its principles and the evidence that supports them. Techniques spread more rapidly than evidence, because the evidence is often locked in dense, inaccessible, inconclusive articles. Often, teachers receive

> a description of what to do and how to do it, but no description of why it might work. There is no explanation of the underlying learning principles on which the method or resources have been constructed. The result is that teachers are constantly being

> encouraged to try out new ideas or methods without understanding how they might be affecting student learning. . . . Unless you have a good understanding of how the technique or resource is supposed to affect student learning, your adaptions can only be trial and error.
>
> (Nuthall, 2007, p. 14)

Trial and error is a slow way for teachers to learn; it is unlikely to lead to the discovery of counter-intuitive findings, such as ways that worse student performance initially leads to better subsequent retention (see Problem 4). Yet, unless we understand the underlying principles, we may imitate the form of a technique but lose sight of its purpose (see, for example, Yin et al., 2008). Worse, we may create "lethal mutations," adapting techniques so radically that the connection to the principle is lost entirely (Wiliam, 2016, p. 173): students copying lesson objectives into books, for example. *Responsive Teaching* introduces evidence around each problem and the rationale for the principles suggested; it shares, not 'what works', but the theoretical underpinnings of promising approaches (Baird et al., 2017). There are no 'responsive teaching techniques': teaching is responsive when teachers use the evidence to identify effective ways to address these problems. Using mini-whiteboards is not responsive teaching, for example; using mini-whiteboards to identify what students are thinking, and adapting accordingly, is. This book seeks to articulate clear principles and the evidence underlying them.

Nevertheless, knowing the evidence and the principles is insufficient,

> teachers will not take up attractive sounding ideas, albeit based on extensive research, if these are presented as general principles which leave entirely to them the task of translating them into everyday practice.
>
> (Black and Wiliam, 1998b, pp. 15–16)

Research offers guidance about what works in laboratories and carefully controlled trials in schools; occasionally, it provides models. However, research will never tell teachers exactly what to do (Wiliam, 2016, p. 98). Researchers cannot prescribe how to apply principles to a specific lesson; only a teacher can turn principles into practices which check Year 9's understanding of balancing equations on a wet Friday afternoon. The evidence has limits; few classroom techniques have been tested rigorously. Teachers therefore need

> a variety of living examples of implementation, by teachers with whom they can identify and from whom they can both derive conviction and confidence that they can do better, and see concrete examples of what doing better means in practice.
>
> (Black and Wiliam, 1998b, p. 16)

This book shows how teaching can be adapted using the principles of responsive teaching; it shares teachers' descriptions of their struggles to make these principles fit their classrooms. Ultimately,

> principles are powerful but cases are memorable. Only in the continued interaction between principles and cases can practitioners and their mentors avoid the inherent

> limitations of theory-without-practice or the equally serious restrictions of vivid practice without the mirror of principle.
>
> (Shulman, 1996, cited in Darling-Hammond et al., 2005, p. 430)

While the implementation of Assessment for Learning proved "top-down and directive" (Swaffield, 2009, p. 11), responsive teaching is done by teachers, taking the principles, examining the examples and adapting to meet their students' needs. We will only achieve the "substantial rewards of which the evidence holds out promise . . . if each teacher finds his or her own ways of incorporating the lessons and ideas . . . into his or her own patterns of classroom work" (Black and Wiliam, 1998b, p. 15). Techniques do not work universally; teachers need to "adapt, adjust and make appropriate professional judgments" (Loughran, Berry and Mulhall, 2012, p. 2). Responsive teaching is expert teaching: choosing approaches which apply principles to suit specific classrooms and students.

Each chapter focuses on one problem and is structured to link underlying principles with classroom examples:

- **Problem:** I frame the problem to set out the aim and why it matters, describing either my own errors or colleagues' struggles.
- **Evidence:** I introduce evidence about ways the problem can be addressed, using studies from schools, laboratories and outside education; this establishes the rationale for the principles and practices and offers a summary for the busy and a reading list for the curious. (I also recommend one key paper which has informed the formulation of each chapter).
- **Principle:** The evidence shapes a principle which can be applied to any class, subject and school.
- **Practice:** I exemplify the principle, showing how weak approaches can be refined to incorporate the evidence. If the weak approaches seem implausible, I can only assure you they represent traps I have fallen into, and reassure you that, in finding them implausible, you have grasped the point I hoped to make.
- **Experiences:** Describing effective practices can imply that change is straightforward; teachers' experiences offer guidance and show how improving teaching is messy, gradual and iterative.
- **Checklists:** I summarise the response to the problem in a checklist to help teachers recall and prioritise critical steps under pressure.

The final chapter suggests how the ideas in the preceding six chapters can be applied by teachers in specific roles; sub-sections are tailored to the needs of new teachers, mentors, leaders and so on.

Responsive teaching does not demand working harder; it invites working differently. Teachers are working hard already, planning, assessing and marking. The approaches in this book suggest tweaks which may allow more effective and efficient planning, assessment and marking. Changing practice demands an initial investment of time and effort; this should pay dividends, increasing student learning and, in due course, autonomy. Better student learning should free teachers to support students in new ways, or to go home earlier. Assessing student learning accurately and tailoring support for them should make teachers more confident and more satisfied (Ross and Bruce, 2007). This book advocates working differently, not working harder.

While books like *Responsive Teaching* could be written for every subject and phase, I am not convinced they yet exist. I have drawn on rich veins of literature from numerous subjects and sought to articulate the underlying principles which transcend them. The examples show how these principles apply across subjects and phases: we can track student thinking and adapt teaching at any age; we can use hinge questions to do so in any subject. Putting this into practice is not just specific to the subject, it is specific to the topic: likely misconceptions and potent questions differ between *The Tempest* and *All's Well that Ends Well*. The underlying principles and the process of improvement transcend subjects. I have tried to provide the tools needed for teachers to apply the principles to whatever they are teaching.

The principles of responsive teaching should help students directly, but they should offer other benefits, too. For example, showing students what success looks like and providing feedback is worthwhile; it also supports metacognition, students' capacity to monitor their thinking and adapt their learning accordingly (Casselman and Atwood, 2017; Koriat, 2007). Tests show what students remember; they also improve students' recall of what is tested (Pashler et al., 2007). There are few ways to support students with low prior attainment to catch up with their peers: students with more existing knowledge learn more and more easily (Willingham, 2009, p. 44). Yet some approaches described in this book have helped all students improve, and helped those with low prior attainment improve most; these include showing students what success looks like, explaining why we are offering feedback and providing comments rather than grades. They have also increased students' motivation and belief they can succeed (White and Frederiksen, 1998; Yeager et al., 2014; Butler, 1988). Formative assessment provides some of the conditions to enter a state of flow, the optimal experience in which we are challenged at just the right level and receive frequent feedback on our performance (Csikszentmihalyi, 2002). Responsive teaching is no panacea, but it offers immediate benefits for student learning while contributing to other important goals.

Conclusion

I am a teacher, and have had the fortune to spend time reading and supporting fellow teachers; I am not an academic. This book draws on extensive research in fields including assessment, cognitive science, behavioural psychology and deliberate practice. Mastering every field is impossible; for example, a bibliography of papers on students' conceptions of science includes 8400 relevant papers (Duit, 2009); I have read only a handful of these. If an expert in any one discipline finds fault, I ask them to take such faults with a generosity of spirit, and not to hesitate to suggest improvements.

Responsive teaching is a path to improvement. Teachers improve throughout their careers, but the most dramatic gains are in the first three years (Kini and Podolsky, 2016). Many things obstruct our development. Where aims are ambiguous, success is hard to measure and we are in it for the love, it is hard to set goals or monitor progress (Lipsky, 1980). We struggle to identify good learning (Coe, 2013) or good teaching (Strong, Gargani and Hacifazlioğlu, 2011) and are prey to 'motivated reasoning', telling ourselves we have done the best we could have done under the circumstances (Lipsky, 1980; see also Ariely, 2013). These barriers obstruct our efforts to improve.

Being clear as to what we want students to learn, and confronting ourselves with evidence as to whether or not they have done so, offers a path to humility and improvement. Our problematic summative assessments led me to discount evidence that my students were struggling; I attributed their lack of apparent progress to flaws in the assessments, oblivious to the weaknesses in my teaching. Richard Feynman was right to say that "the first principle is that you must not fool yourself – and you are the easiest person to fool" (Feynman, 1974). I hope, as did Benjamin Bloom, that

> when teachers are helped to secure a more accurate picture of their own teaching methods and styles of interaction with their students, they will increasingly be able to provide more favorable learning conditions for more of their students.
>
> (1984, p. 11)

Constantly seeking to improve by identifying what students have learned and responding accordingly should be central to our identity as teachers. Tracking student learning shifts our perspective away from the effort we put into our teaching and towards how well it is working. This has powerful effects: experienced mentors see formative assessment as central to helping teachers improve by focusing on students' needs (Athanases and Achinstein, 2003). Responsive teaching allows us to focus on improving our teaching and our students' learning.

Great reads on this are . . .

Christodoulou, D. (2017). *Making good progress: The future of Assessment for Learning*. Oxford: Oxford University Press.

Daisy Christodoulou dismantles formative and summative assessment to show their distinct purposes and suggest ways to apply them more effectively.

Willingham, D. (2009). *Why don't students like school? A cognitive scientist answers questions about how the mind works and what it means for the classroom*. San Francisco, CA: Jossey-Bass.

Daniel Willingham summarises how we think and learn clearly, succinctly and useably. He describes key experiments, shows the psychological principles at work and demonstrates how teachers can apply them.

Note

1 Technically, there is no such thing as a 'summative' assessment; it is the inferences drawn from the assessment, which is 'summative' or 'formative'. In practice, the term is so well used, and so much less clunky than 'assessments for summative purposes', that I have used it on occasion.

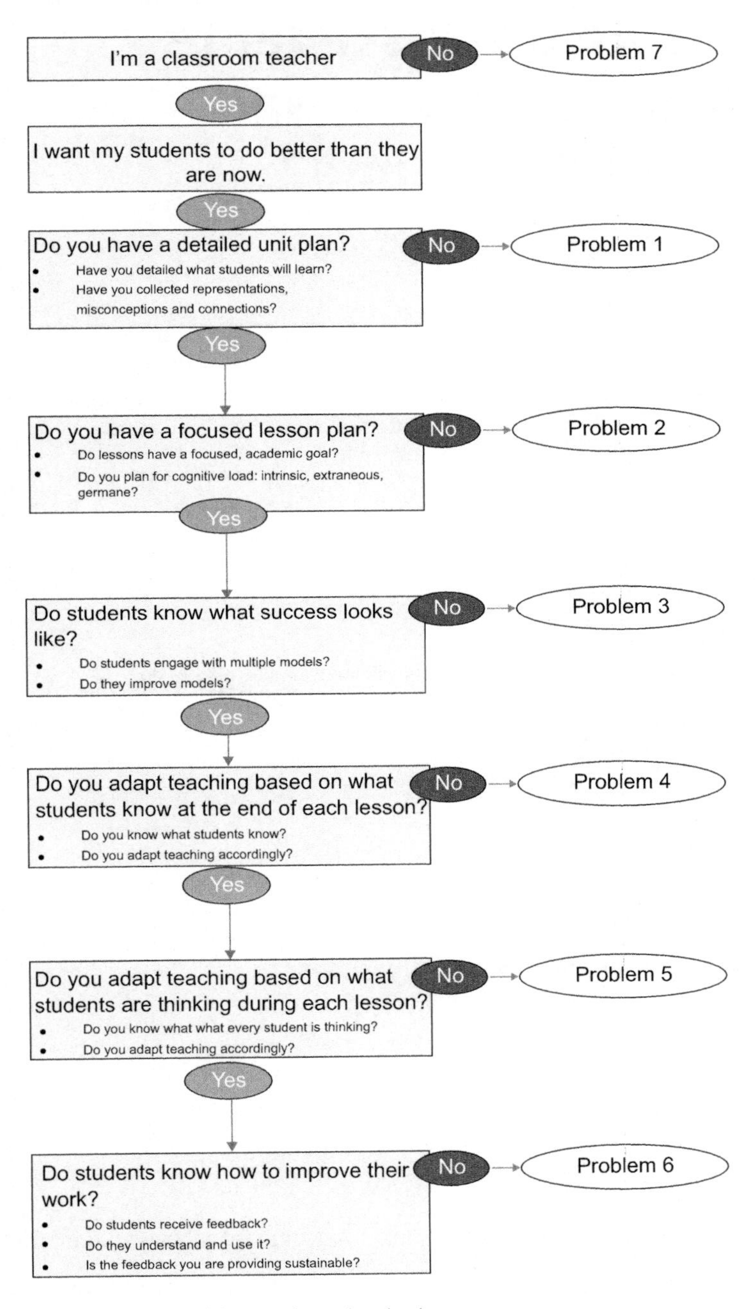

Figure 0.1 Responsive teaching – where to start

1 How can we plan a unit, when we want students to learn so much, and have so little time?

The problem
There is so much to teach and not enough time in which to teach it; we cannot respond to student learning until we are sure what matters most.

The evidence
Focus on the most powerful knowledge
Specify what students are to learn
Identify connections between ideas
Plan for units, not lessons

The principle
Responsive teachers specify what students will know and be able to do

Practical tools
Knowledge organisers
Pedagogical content knowledge unit planners

- Representations
- Misconceptions
- Horizon knowledge
- Sequence

Experience - Emma McCrea/Marcus Bennison
Planning subject knowledge for teaching

Checklist

The problem

There is so much to teach, and not enough time in which to teach it: we cannot respond to student learning until we are sure what matters most.

Maya has been teaching the class for eighteen months when she realises she's in trouble. She has taught as fast as she can, faster than she'd like, and yet she's still only covered around half of what she wants students to learn. Students are expected to know so much more with the new specifications, and Maya wants to teach more than just the specification. Even what she's covered doesn't seem to be sinking in. She tests her students and finds that much of what she taught last year has evaporated; absentees struggle, having missed their one chance to grasp a topic. Planning takes ages, as Maya struggles to establish what matters most from the syllabus and find ways to explain it. Despite her planning, what students recall at the end of a unit feels like a lottery based on how each lesson went. As soon as one lesson is done, Maya begins preparing the next, an exhausting treadmill, and while Maya and her colleagues share lesson plans and resources, Maya rarely finds this useful.

Maya wants to respond to students' needs, but she is spending all her time struggling to plan effectively. She cannot assess what students have learned, and respond, until she is clear about what she most wants them to learn. Maya concludes that she needs to begin by reviewing how she approaches planning and identify:

- What should she prioritise?
- How can she create detailed, flexible plans?
- How can she make these plans useful for colleagues?

The evidence

Focus on the most powerful knowledge
Specify what students are to learn
Identify connections between ideas
Plan for units, not lessons

Graham Nuthall (2007) examined student learning minutely, tracking, for example, how an individual student's confusion of refraction, reflection and magnification developed with every new task across a unit of learning, rendering much of the information she was exposed to meaningless. He reached a deceptively simple conclusion:

> We discovered that a student needs to encounter, on at least three different occasions, the complete set of the information she or he needed to understand a concept. If the information was incomplete, or not experienced on three different occasions, the student did not learn the concept.
>
> (p. 63)

Therefore, he argued:

> Student learning primarily depends on the information they are exposed to. This means that activities need careful designing so that students cannot avoid interacting with this relevant information. It also means being very careful about the form of the information that is encountered.
>
> (p. 79)

Maya realises she must convert her general ambitions for students into specific goals and design repeated opportunities to meet these goals. Her role is not curriculum design, but planning teaching which will allow students to learn the concepts set out in broad terms by the curriculum (Young, 2014a, pp. 94–97). Her reading of the evidence convinces her that her planning should be specific and knowledge-focused, should connect ideas and should be shaped around units, rather than lessons.

1. Focus on powerful knowledge

Maya has to decide the most important things for students to learn. She finds Anna Sfard's (1998) discussion of competing metaphors of learning helpful. Learning may be seen as:

- **Acquisition:** gaining "basic units of knowledge that can be accumulated, gradually refined, and combined to form ever richer cognitive structures" (p. 5).
- **Participation:** becoming a member of a community, able to "communicate in the language of this community and act according to its particular norms" (p. 6).

This seems to explain many differences between teachers, from the purpose of education to the best activity in a lesson. Teachers focused on acquisition might see teacher explanation as an efficient way for students to gain knowledge; those focused on participation might see students' pursuit of the scientific method as more important than their reaching correct conclusions today. Sfard suggests that the two metaphors are not exclusive: Maya comes to believe that acquiring knowledge and using it with increased flexibility allows students to participate by introducing them to communities of specialists (Young, 2014a, p. 101). An apprentice carpenter would not be left to trial and error among peers – their acquisition of knowledge and skill through increasingly challenging tasks culminates in the creation of a masterpiece, demonstrating their learning and qualifying them to participate as a carpenter. Acquisition and participation rely on students gradually developing fluency, automating simple procedures and developing more complicated mental models. Maya concludes that acquiring knowledge will qualify students to participate in the domains they wish to master; this requires identifying which knowledge matters most.

With limited time, Maya prioritises powerful knowledge, which will help her students understand the world and study, work and live as they wish (Young, 2014b). This is organised within subjects and:

1 **Has explanatory power:** knowledge of the Ancient World helps students understand literature, art and music; the ability to add integers prepares students to add fractions and decimals and to multiply.
2 **Is unlikely to be encountered outside school:** technical terms and classic literature are less likely to be encountered outside school than everyday language and young adult fiction.

Maya views powerful knowledge critically; it is not perfect and not fixed, and the choices and motives of those selecting it can be questioned (Young, 2014b). She recognises the value and the limits of the canon, and ensures the inclusion of women authors and mathematicians from outside the European tradition. Teaching powerful knowledge is not the sole solution to social problems, but Maya is convinced that all students are entitled to this knowledge and the understanding it allows (Young, 2014b). Learning familiar topics can be beneficial and enjoyable, but Maya uses them as intermediate steps to educate her students, broadening their horizons by leading them beyond what they already know.

- **Maya prioritises teaching the most powerful knowledge.**

2. *Specify what students should learn*

Maya finds that effective planning must be surprisingly specific. Vague objectives are common in education (Millar, 2016). One example sticks with Maya: "Can compare two fractions to identify the larger" sounds specific, but students' success depends on the fractions selected:

- $\frac{3}{7}$ and $\frac{5}{7}$ — 90% of students were correct.
- $\frac{3}{4}$ and $\frac{4}{5}$ — 75% of students were correct.
- $\frac{5}{7}$ and $\frac{5}{9}$ — 15% of students were correct.

(Hart, 1981, in Wiliam, 2010, pp. 254-255)

Even a seemingly precise objective does not establish what students should be able to do. A clear standard would specify which fractions students will master: those with shared denominators, different denominators or different and unusual denominators. Similarly, if Maya is to teach Newton's Second Law of Motion, that force is mass multiplied by acceleration, does this means students should memorise the formula? Will they be told to apply it to data or be given a scenario and expected to recognise that it applies? Should they be able to use it to justify lower speed limits? Should they know the bounds of the law under special relativity? (Parkes and Zimmaro, 2016, p. 47). Some subjects have more clearly specified bodies of knowledge than others, but whether Maya is teaching Viking invasions, irregular verbs or *Oliver Twist*, she needs to decide what she expects students to learn if she is to plan with confidence. Maya resolves to identify specific goals which will help her allocate time and design assessments.

Maya is teaching a curriculum of concepts, ideas and connections, but she needs to specify the ideas which will help students learn those concepts if they are to make sense of the curriculum. Specifying knowledge feels reductive, but Maya sees its importance and value. A secondary school history teacher may have only one hundred hours to teach the entirety of human history before students can drop the subject. All teachers have similar choices to make, squeezing vast domains of knowledge into limited time. Maya prefers to prioritise explicitly, transparently and collectively, rather than tacitly, spontaneously and individually. Specifying what students are to learn also refreshes her subject knowledge and prepares her to share what success looks like (Problem 3).

- **Maya specifies the basic ideas she wants students to learn in order to understand the underlying concepts.**

3. Identify connections and threshold concepts

Gaining specific knowledge is not enough, however: Maya wants students to use this knowledge analytically, critically and creatively. Previously, she struggled because she was not clear what she wanted students to learn; now that she is clear about the knowledge underpinning learning, she does not want to overlook the importance of students organising, connecting and applying their knowledge. She recognises that students cannot have deep knowledge of everything, and that shallow knowledge is better than nothing (Willingham, 2009, p. 49). Nevertheless, she wants students to convert isolated elements of knowledge into connected, useful mental models. Maya plans to help students use their factual knowledge to do three things:

1. **Make connections:** Maya wants her students to make connections between topics within the subject, between years and across subjects, helping them to apply what they have learned and to make sense of the curriculum as a whole.
2. **Develop knowledge of substantive concepts:** Maya wants students' factual knowledge to develop into an understanding of substantive concepts. She wants them to recognise that words like force, source and monarch mean different things in different contexts. Maya wants her students to be able to follow historians as they shift between using 'church' to mean a building and an institution, e.g.: "Changes to church doctrine during the Reformation led to substantial alterations to church decoration."
3. **Pass through threshold concepts:** Maya wants students to enjoy lightbulb moments which change the way they view the subject, and the world. Threshold concepts seem a useful framework for this. They are:
   - Troublesome: difficult to understand
   - Transformative: of a student's perspective
   - Irreversible: once learned, they are hard to unlearn
   - Integrative: they show how different ideas are related
   - Bounded: there are limits to the insight they offer (Meyer and Land, 2003)

   Maya sees huge power in threshold concepts: students view the world differently when they recognise that a narrator may be unreliable, a king may depend on his barons and that the stronger the forces holding particles together, the more energy is required to overcome them. Threshold concepts are individual realisations, but Maya realises she can plan to introduce ideas and invite reflection to promote these realisations.

- **Maya plans the connections, threshold concepts and substantive concepts she hopes students will grasp.**

4. Plan units, not lessons

Maya realises that while she talks about 'planning lessons' more often than 'planning units', it is more useful to plan units: sequences of lessons on forces, the English Civil War or Act II of *Hamlet*. Building on the school's curriculum, Maya finds that unit planning allows:

- **Repetition:** learning is the gradual acquisition, reorganisation and refinement of knowledge, not a one-off event. Student performance while being taught is a poor indicator of lasting learning (Soderstrom and Bjork, 2015); spaced revisiting of key ideas is critical

(Pashler et al., 2007). Planning units allows Maya to revisit key ideas, increasing students' chances of understanding and helping those who have been absent.

- **Coherence:** Maya is not teaching isolated ideas; a unit plan helps her create coherent sequences which form a narrative that students can structure and recall (Gentner, 1976; Willingham, 2004). In teaching a text in English, for example, she wants students to have a strong factual knowledge of the text, but also to connect context, quotations, big ideas and the author's purpose in their arguments (Tharby, 2017). Planning units allows Maya to foreshadow, highlight and reiterate connections between ideas and tackle misconceptions across units.
- **Ease:** Planning units allows Maya to identify how students will learn key ideas in advance. She can plan lessons easily based on the unit plan and knowledge of what students have learned so far (Problem 4).

- **Maya focuses on planning units, not lessons.**

The principle

Responsive teachers specify what students will know and be able to do

Practical tools

Knowledge organisers
Pedagogical content knowledge organisers

- Representations
- Misconceptions
- Connections
- Sequence

Based on the curriculum, Maya wants a unit plan which:

- Specifies powerful knowledge and crucial vocabulary
- Designs progression, repetition and coherence within the unit
- Supports her teaching by including:
  - Ways to explain and represent key ideas
  - Misconceptions
  - Opportunities for connections and exposure to threshold concepts

Version 1

Maya used to plan each lesson the week she taught it. She would specify what students needed to know - about the effects of the Great Depression on Germany, for example. She chose objectives (Problem 2), designed exit tickets (Problem 4) and constructed activities towards them. Lessons went well, but they were isolated. Students understood topics better by linking them to their prior knowledge: remembering the positions of workers and nobles in Germany made the effects of the Depression on each group more comprehensible. The social consequences of the Depression, discussed in one lesson, made sense of the political consequences discussed in the next. Maya helped students make these connections, but she did so spontaneously, as they occurred to her and her students. Results were uneven; sometimes students reached the end of the unit and Maya found the opportunity for a connection had been lost, or that a key idea had fallen into the cracks between lessons. Designing individual lessons kept Maya so busy that she struggled to prepare for these connections and the revisiting of key ideas. Students remembered and understood far less than she hoped. Maya wanted to move from planning lessons to planning units.

Version 2

Maya specified exactly what she wanted students to learn during each unit. The lesson on the Great Depression was part of a unit on Hitler's rise to power: Maya created a knowledge organiser which collated everything students had to know to understand the underlying ideas (Figure 1.1). She created a timeline of key events, listed key characters and key terms and the factors in Hitler's rise to power that mattered most. She wrote no more than one sentence in each category and cut everything that was not concrete and essential.

Maya considered mentioning the skills students would develop through the unit - in this case, explaining the causes of Hitler's rise to power. She resisted this urge, however, because she wanted students to be able to explain both the causes of the events and the change and continuity they demonstrated. She found that knowledge organisers worked well with different ages and subjects, but with some variations in form (Figures 1.2 and 1.3). In subjects such as science and geography, colleagues often included diagrams. In literature, themes seemed to replace the key factors Maya included. Teachers chose different ways to organise the knowledge - sometimes by hierarchy, sometimes by association - but each found a different way to make it fit their subject.

	The Nazi rise to power and seizure of control
1929	**Wall Street Crash** in America leads to a depression and **6 million unemployed** in Germany; Nazi messages seem more relevant.
1930	Election: Nazis win **18.3%** of votes: **second largest party** in Reichstag.
	Weak centre-right governments are supported by Hindenburg.
1932 Apr	Presidential election: **Hindenburg wins again**, Hitler comes second.
Jul	Election: Nazis gain **37.4%** of votes but no government is formed.
Nov	Election: Nazi vote falls to **33.1%**, Communist vote increases; Kurt von Schleicher appointed chancellor but can't gain support
1933 Jan	Von Papen convinces Hindenburg to **appoint Hitler chancellor** as part of a **coalition government** which will limit Nazi power.
Feb	The **Reichstag Fire; Reichstag Fire Decree** restricts civil liberties.
Mar	Election: Extensive intimidation by Nazis; Nazis win **44%** of votes, **ban the Communist Party,** pass the **Enabling Act**.
May	**Trade unions** are **banned.**
Jul	All other **political parties are banned.**
1934 Jun	**Night of the Long Knives:** Hitler curbs the power of the **SA, Röhm is killed,** leading **opponents arrested** including von Schleicher.
Aug	Hindenburg dies, **Hitler becomes Führer.** Armed forces swear a personal **oath of loyalty** to Hitler.

Key factors	
1. Unemployment	8 million unemployed people felt let down by the government.
2. Propaganda	The Nazis offered appealing messages through powerful propaganda, Hitler was an effective speaker.
3. Terror	The SA attacked opposing politicians and supporters; violence created an atmosphere of crisis Hitler promised to solve.
4. Fear of Communism	Unemployment and anger increased votes for the Communists: this scared many middle-class voters.
5. Uncommitted democrats	Hindenburg and many right-wing non-Nazis were not committed to democracy and underestimated Hitler.

Key terms	
1. Chancellor	Head of government
2. Coalition	Government with two or more parties sharing power
3. Enabling Act	Law giving Hitler power to make laws without Reichstag approval.
4. Führer	The combination of chancellor and president
5. President	Head of state
6. Reichstag	The German national parliament
7. SA	Organisation of Nazi supporters, 2 million members by late 1933.
8. SS	Hitler's elite bodyguard

Key people	
1. Paul von Hindenburg	President, First World War general and hero
2. Franz von Papen	Right-wing member of the Centre Party
3. Ernst Röhm	Leader of the SA
4. Kurt von Schleicher	Right-wing politician, former general, not a Nazi

Figure 1.1 Knowledge organiser: Hitler's rise to power

Crime and Punishment | Year Five | Autumn 2

Timeline – Development of Law

1	c. 2100 BCE	The **Sumerians** produce the earliest known, **written** record of laws.
2	c. 1790 BCE	Babylonian legal codes based on law derived from the **will of the Gods**.
3	c. 130-180 CE	Romans systemize law, but still consider crimes as 'wrongs' between two **individuals**.
4	c. 1100 CE	In Britain the idea of committing a wrong against '**State**' emerges.
5	c. 1100-1200 CE	Britain starts using courts to trial people. The decisions become **common law**.
6	1215	The **Magna Carta** guarantees individual rights and brings the King under the law.
7	1651	**Thomas Hobbes** argues for a powerful state to enforce laws in **Leviathan**.
8	1762	**Jean-Jacques Rousseau** argues people are 'noble savages' in **The Social Contract.**
9	1948	Universal Declaration of Human Rights guarantees right to a fair trial and no punishment without law.
10	1965	Britain abolishes the death penalty for murder.

Vocabulary

1	crime	An action that deliberately **harms** either a person, their property or a community.
2	morality	A system to decide if something is **right or wrong.**
3	court	A meeting to **decide** whether someone is guilty of breaking a law.
4	judge	The **person** who is in **charge** of the court and makes decisions about the law.
5	sentence	The **punishment** that is given to someone for breaking the law.
6	dominion	The act of owning and controlling property (which could include people: slaves)
7	treason	The crime of betraying one's country, especially by attempting to kill the monarch or overthrow the government.
8	torture	Inflicting extreme and severe pain on someone, sometimes to force them to do or say something.
9	capital punishment	The death penalty.
10	jury	A group of ordinary people who decide if a person is guilty or innocent.

Key ideas

1	Natural law	The idea that all people have certain rights, and can work out what is right or wrong through using reason.
2	War of all against all	Hobbes's idea of how the world would be if there was nobody in charge.
3	Noble savage	The idea that people are naturally good, but make poor choices when forced to live together.
4	Presumption of innocence	The idea that everyone should be considered innocent until proven guilty.
5	Common law	Decisions of judges in particular cases which then get followed by other judges.
6	Sin	An act that is considered to be against what God wants.
7	Policing	The state gives certain people the power to arrest and detain people: police officers.
8	Social contract	The idea that all of us agree to behave in a certain way, so that everyone is safe and can be successful.
9	Laws	A system of rules that are decided by the government and enforced by the police and courts.
10	Juvenile court	A special court, which decides the innocence or guilt of children who have committed crimes.

Punishments through the ages

1	Banishment	Being sent away from the country or place you live for either a fixed time or forever. In the olden times this would make life very difficult, as travelling was difficult and they may not be accepted to other places.
2	Execution	Being killed. Common executions included beheading, hanging, firing squad, electric chair and lethal injection.
3	Corporal punishment	Until very recently children were allowed to physically hurt children as a punishment. Common punishments included caning (hitting children with a cane on the hand, legs or rear. Sometimes a hand, a belt or a slipper would be used.
4	Flogging	People are whipped with rope or with sticks, usually across their bare back or the bare soles of the feet.
5	Stocks	People's hands and head are locked into a wooden block, usually in a public area to embarrass them.
6	Hard labour	Prisoners are forced to do very difficult, exhausting work, like breaking rocks, or building roads.
7	Imprisonment	When someone is put in prison, or jail, for a set amount of time (sometimes forever).

Figure 1.2 Year 5 knowledge organiser: *Crime and Punishment*

Courtesy of Jon Brunskill

Chapter	Plot
1 The Story of the Door	*Passing a strange-looking door whilst out for a walk, Enfield tells Utterson about incident involving a man (Hyde) trampling on a young girl. The man paid the girl compensation. Enfield says the man had a key to the door (which leads to Dr Jekyll's laboratory).*
2 Search for Hyde	*Utterson looks at Dr Jekyll's will and discovers that he has left his possessions to Mr Hyde in the event of his disappearance. Utterson watches the door and sees Hyde unlock it, then goes to warn Jekyll. Jekyll isn't in, but Poole tells him that the servants have been told to obey Hyde.*
3 Dr Jekyll was Quite at Ease	*Two weeks later, Utterson goes to a dinner party at Jekyll's house and tells him about his concerns. Jekyll laughs off his worries.*
4 The Carew Murder Case	*Nearly a year later, an elderly gentleman is murdered in the street by Hyde. A letter to Utterson is found on the body. Utterson recognises the murder weapon as a broken walking cane of Jekyll's. He takes the police to Jekyll's house to find Hyde, but are told he hasn't been there for two months. They find the other half of the cane and signs of a quick exit.*
5 Incident of the Letter	*Utterson goes to Jekyll's house and finds him 'looking deadly sick'. He asks about Hyde but Jekyll shows him a letter that says he won't be back. Utterson believes the letter has been forged by Jekyll to cover for Hyde.*
6 Remarkable Incident of Dr Lanyon	*Hyde has disappeared and Jekyll seems more happy and sociable until a sudden depression strikes him. Utterson visits Dr Lanyon on his death-bed, who hints that Jekyll is the cause of his illness. Utterson writes to Jekyll and receives a reply that suggests he is has fallen 'under a dark influence'. Lanyon dies and leaves a note for Utterson to open after the death or disappearance of Jekyll. Utterson tries to revisit Jekyll but is told by Poole that he is living in isolation.*
7 Incident at the Window	*Utterson and Enfield are out for walk and pass Jekyll's window, where they see him confined like a prisoner. Utterson calls out and Jekyll's face has a look of 'abject terror and despair'. Shocked, Utterson and Enfield leave.*
8 The Last Night	*Poole visits Utterson and asks him to come to Jekyll's house. The door to the laboratory is locked and the voice inside sounds like Hyde. Poole says that the voice has been asking for days for a chemical to be brought, but has rejected it each time as it is not pure. They break down the door and find a twitching body with a vial in its hands. There is also a will which leaves everything to Utterson and a package containing Jekyll's confession and a letter asking Utterson to read Lanyon's letter.*
9 Dr Lanyon's Narrative	*The contents of Lanyon's letter tells of how he received a letter from Jekyll asking him to collect chemicals, a vial and notebook from Jekyll's laboratory and give it to a man who would call at midnight. A grotesque man arrives and drinks the potion which transforms him into Jekyll, causing Lanyon to fall ill.*
10 Henry Jekyll's Full Statement of the Case	*Jekyll tells the story of how he turned into Hyde. It began as a scientific investigation into the duality of human nature and an attempt to destroy his 'darker self'. Eventually he became addicted to being Hyde, who increasingly took over and destroyed him.*

Character	
Dr Henry Jekyll	*A doctor and experimental scientist who is both wealthy and respectable.*
Mr Edward Hyde	*A small, violent and unpleasant-looking man; an unrepentant criminal.*
Gabriel Utterson	*A calm and rational lawyer and friend of Jekyll.*
Dr Hastie Lanyon	*A conventional and respectable doctor and former friend of Jekyll.*
Richard Enfield	*A distant relative of Utterson and well-known man about town.*
Poole	*Jekyll's manservant.*
Sir Danvers Carew	*A distinguished gentlemen who is beaten to death by Hyde.*
Mr Guest	*Utterson's secretary and handwriting expert.*

Themes
The duality of human nature
Science and the unexplained
The supernatural
Reputation
Rationality
Urban terror
Secrecy and silence

Vocabulary
aberration
abhorrent
allegory
allusion
anxiety
atavism
consciousness
debased
degenerate
depraved
duality
duplicity
epistolary
ethics
eugenics
feral
genre
metamorphosis
perversion
professional
respectability
restraint
savage
subconscious
suppression
supernatural
unorthodox
Victorian

Context
Fin-de-siècle fears – at the end of the 19th century, there were growing fears about: migration and the threats of disease; sexuality and promiscuity; moral degeneration and decadence.
Victorian values – from the 1850s to the turn of the century, British society outwardly displayed values of sexual restraint, low tolerance of crime, religious morality and a strict social code of conduct.
The implications of ***Darwinism and evolution*** haunted Victorian society. The idea that humans evolved from apes and amphibians led to worries about our lineage and about humanity's reversion to these primitive states.
Physiognomy – Italian criminologist Cesare Lombroso (1835-1909) theorised that the 'born criminal' could be recognised by physical characteristics, such as asymmetrical facial features, long arms or a sloping forehead.
Victorian London – the population of 1 million in 1800 to 6.7 million in 1900, with a huge numbers migrating from Europe. It became the biggest city in the world and a global capital for politics, finance and trade. The city grew wealthy.
Urban terror – as London grew wealthy, so poverty in the city also grew. The overcrowded city became rife with crime. The crowd as something that could hide sinister individuals became a trope of Gothic and detective literature.
Robert Louis Stevenson was born and raised in Edinburgh, giving him the dual identity of being both Scottish and British. Edinburgh was a city of two sides - he was raised in the wealthy New Town area, but spent his youth exploring the darker, more sinister side of town.
Deacon Brodie – a respectable member of Edinburgh's society and town councilor, William Brodie led a secret life as a burglar, womaniser and gambler. He was hanged in 1788 for his crimes. As a youth, Stevenson wrote a play about him.

Figure 1.3 GCSE knowledge organiser: Jekyll and Hyde

Courtesy of James Theobald

Maya liked using knowledge organisers, but they did not solve her problems entirely. They summarised the unit comprehensibly and helped her plan more focused lessons; she chose an appropriate segment of the knowledge organiser to teach and revisited key ideas across lessons. Specifying key terms allowed her to introduce them strategically and check students' understanding of them across the unit. Maya found knowledge organisers a useful practical tool. With minimal changes, they provided a homework, cover lesson, revision timetable and test of prior knowledge; she could create a quiz in sixty seconds, deleting the events and asking students to recreate the timeline or deleting the key terms and asking students to recall them from their definitions. Nonetheless, Maya realised that although it helped her plan what students were to learn, it did not prepare her to teach or to make connections between ideas: she was still making rapid decisions between lessons. She wondered if she could plan key features of her teaching in advance, while maintaining the flexibility to respond to students.

Version 3

Maya decided that planning key features of her teaching would require her focusing on pedagogical content knowledge:

> The blending of content and pedagogy into an understanding of how particular topics, problems, or issues are organized, represented, and adapted to the diverse interests and abilities of learners, and presented for instruction.
>
> (Shulman, 1987, p. 8)

Maya found the extensions to this definition offered by Ball, Thames and Phelps (2008) a useful framework for her planning:

- Knowledge of content and teaching
- Knowledge of content and students
- Horizon knowledge

1. Knowledge of content and teaching: representations and explanations

> Teachers . . . choose which examples to start with and which examples to use to take students deeper into the content. Teachers evaluate the instructional advantages and disadvantages of representations used to teach a specific idea and identify what different methods and procedures afford instructionally.
>
> (Ball, Thames and Phelps, 2008, p. 401)

In planning, Maya selects representations and explanations which help students get closer to understanding the true complication of an idea. These include:

- **Images:** photos from performances of *Othello* to examine character, islands of plastic at sea to show the effects of pollution, portraits of historical figures to examine how they presented themselves.

- **Diagrams:** electrical circuits, the circulatory system, the creation of an ox-bow lake.
- **Graphs:** population to understand demography, election results to understand changes in power, the effect of temperature on photosynthesis.
- **Examples:** a quotation from *Hamlet* encapsulating his character, washing powder showing the action of enzymes, specific problems to show mathematical procedures.
- **Stories:** individuals' journeys exemplifying migration trends, the life of a Sikh as an introduction to the religion, individuals' changing employment during the Industrial Revolution.
- **Experiments:** growing plants in the classroom to understand photosynthesis, investigating different ways students can make the same number to understand addition.
- **Analogies:** the school's organisational hierarchy likened to the structure of government, students in the playground representing the solar system, the circulatory system brought to life by students passing oxygen to one another.

Maya has always used representations, but previously, she selected them before each lesson; creating a bank of options before the unit has several advantages. It allows Maya to weigh the advantages and disadvantages of each representation; for example, using a lift's journey up and down a building represents addition and subtraction as positional (moving up and down) but does not allow students to make sense of 4 - (-3) (Ball, 1993, p. 9). Students need repeated exposure to key ideas, but this

> does not mean simple repetition . . . [which] is likely to be boring . . . What it does seem to mean is that students' minds need time to process new information. They need opportunities to come at the material in different ways.
>
> (Nuthall, 2007, p. 81)

A bank of options allows Maya to introduce ideas with the representation which seems likely to help students most, and offers her alternatives if students struggle. It allows her to link concrete and abstract representations and helps students transfer learning from one context to another through 'concreteness fading': introducing an idea with concrete representations, then shifting towards increasing abstraction (Pashler et al., 2007), for example, beginning with photos of a biological process and shifting to diagrams. She believes her collection could be shared and used almost anywhere, irrespective of school policies and practices: a telling image remains a telling image in almost any context.

2. Knowledge of content and students: misconceptions

> Teachers must anticipate what students are likely to think and what they will find confusing. . . . Central to these tasks is knowledge of common student conceptions and misconceptions.
>
> (Ball, Thames and Phelps, 2008, p. 401)

Maya has noticed that experienced colleagues have a sixth sense for existing student misconceptions and those that students develop during a unit. Misconceptions are beliefs which conflict with what is to be learned, rather than errors or knowledge gaps (Chi, 2008), like the belief that objects sink because they are heavy, rather than because they are dense. Misconceptions include:

- **Over-generalising a rule:** because some words ending in 's' require an apostrophe, most words ending in 's' receive one.
- **Simplified views of substantive concepts:** seeing 'the church' literally leaves students believing that before the Reformation, when there was only 'one church', people had to walk miles to reach it.
- **Common sense reactions to the counter-intuitive nature of some academic knowledge:** asked to round 0.01 down to one decimal place, students refuse to round to 0.0, because 0 means nothing, and students began with something.
- **While many misconceptions are universal, some are culturally specific:** in Turkish, to switch a light on, you 'open' it; this causes problems when studying electrical circuits, in which you 'open' a light to break the circuit and switch the light off.

Misconceptions are often students' logical deductions, or teachers' or parents' handy simplifications, but they can hinder learning. Maya wants to be prepared to address them.

Since most misconceptions are specific to subjects and topics, Maya struggles to plan for them before teaching a course for the first time. In science and maths, she can draw on extensive collections of misconceptions; she uses the American Association for the Advancement of Science (n.d.) collection, which tells her that a quarter of Americans aged 11–14 believe that cells are not made up of atoms, a third think that chemical changes are irreversible and over half believe that some organisms do not have DNA. She also finds extensive collections available in maths (Morgan, n.d.) and in chemistry (RSC, n.d.). In subjects without organised collections, Maya plans to draw on her experience and consult colleagues. Maya can prepare responses to misconceptions, too (see Problem 5). People change their conceptual beliefs reluctantly, if at all, even when they encounter contradictory information (Posner et al., 1982), since existing concepts seem to be "protected" against new experiences (Nuthall, 2007, p. 73) and people may maintain misconceptions indefinitely, instead of suppressing them (Potvin, Sauriol and Riopel, 2015). Maya collects ways to challenge misconceptions: she may state that cells are larger than atoms, she may offer an image of an atom and a cell to scale, she may emphasise that all matter is composed of atoms and ask students to explain what cells are made of. Collecting misconceptions makes Maya more alert to them and better prepared to address them.

3. *Horizon knowledge: connections*

> [Horizon knowledge is] an awareness of how mathematical topics are related over the span of mathematics included in the curriculum. First-grade teachers, for example, may need to know how the mathematics they teach is related to the mathematics students will learn in third grade to be able to set the mathematical foundation for what will come later. It also includes the vision useful in seeing connections to much later mathematical ideas.
>
> (Ball, Thames and Phelps, 2008)

Maya wants to help students link current learning with the rest of the curriculum. She challenges them to recall existing knowledge and link it to the current topic, like expert teachers, who frequently begin lessons by asking 'Remember when we learned about . . .?' (Westerman, 1991). Maya wants to foreshadow future connections too: 'When we study Macbeth, you'll see great examples of supernatural influences.' These connections help students reinforce and reorganise their existing knowledge, as well as add to it. Maya wants students to transfer existing learning to new contexts; explicit cues may help them do so (Gick and Holyoak, 1980). Maya hopes these connections will help students respond in more sophisticated ways; breadth of knowledge is implicit in the most sophisticated student answers (Hammond, 2014). Planning connections increases the chances Maya's students will make them.

Maya considers the connections she can make across the curriculum, too. She plans to highlight links between subjects: between science and maths in using graphs, between history and literature when studying Communism and reading *Animal Farm*, between art, PE and maths when looking at use of space and movement. St Matthias School in Bethnal Green have identified vertical, horizontal and diagonal links (Sealy, 2017):

- **Vertical:** connecting concepts across years within the same subject, for example, learning about tyrants in Year 2 (King John), Year 5 (Dionysius of Syracuse) and Year 6 (Hitler).
- **Horizontal:** connecting concepts within the year across subjects, for example, invasions: the Vikings invading England, microbes invading the body and the nominalisation - transformation from verb to noun - of invade into 'invasion'.
- **Diagonal:** connecting concepts across subjects and years, for example, resistance to tyranny by the Jews in Exodus (RE, Year 3), Matilda (literacy, Year 4), and Harriet Tubman (literacy, Year 6).

Repeating key ideas reinforces them for students and develops their understanding of substantive concepts, such as sources of rivers (Year 3), news reports (Year 4) and historical accounts (Year 5). Planning connections allows teachers to highlight them: it can work across an entire school or be done by individual teachers. This approach is powerful because it is interdisciplinary; it asks students to make connections based on their knowledge. Planning to use horizon knowledge helps Maya introduce connections and helps students organise and strengthen their knowledge.

Unit plans

Maya creates unit plans which sequence the knowledge she has specified in the knowledge organiser and the pedagogical content knowledge she has collected. Alongside the main content of the lesson, she notes possible representations and misconceptions, points to foreshadow and revisit connections and threshold concepts students may encounter. The first two lessons in her unit therefore run as seen in Figure 1.4.

Maya also finds that this kind of unit planning could be organised around questions rather than lessons, as demonstrated in Figure 1.5 by Adam Boxer.

Maya uses her unit plan to establish the key ideas to include in lessons; the rest of her lesson planning is quicker, shaped around the key ideas. Her unit plan also allows her to plan to revisit key ideas across the whole year, using tests both to check whether students have retained these ideas and to improve students' retention of them (Pashler et al., 2007).

Lesson topic	Representations	Misconceptions	Revisit	Foreshadow	Connections	Threshold concepts
Great Depression	2008 crash	German money physically in the USA	Hyperinflation in 1923	Increased popular support for the Nazis	Boom years in the USA	
Increased popular support for the Nazis	Propaganda posters Marches: pictures	Everyone/no one supported Hitler	Weaknesses of Weimar democracy	Terror	Extent of popular support for Soviet Russia	Tyranny evokes a range of responses, even from good people

Figure 1.4 A unit sequence

Question	Answer	Explanatory model	Key diagrams	Link to other topics	Misconceptions	Practical opportunity
What is the name of the process of turning a solid into a liquid?	Melting	Most students will know many of these key words already	Flow diagram of movement between ice, water and steam with labelled arrows	Previous questions on states of matter. Should discuss expected properties of each of them as you go	Students may believe that all this only occurs with water and not other substances. Make sure to be clear water is an example	Melting ice – can do as a demo
What is the name of the process of turning a liquid into a gas?	Boiling	Note difference between boiling and evaporating. Evaporating is when it just occurs naturally (technically when phase transition occurs at lower than boiling point)		As above		As above but then to steam
What is the name of the process of turning a gas into a liquid?	Condensing			As above. If sufficient biology has been covered can discuss water leaving the human body		Can discuss breathing onto glass in the winter but could lead to misconception about steam coming out of our bodies at 100°C

Figure 1.5 A unit sequence in chemistry

Courtesy of Adam Boxer

Experience - Emma McCrea/Marcus Bennison

Planning subject knowledge for teaching

Emma McCrea is a senior lecturer in maths education at the University of Brighton, where she specialises in ITT and subject knowledge development. She is an advanced skills teacher, NCETM PD lead, level 3 maths hub lead, CPD trainer for MEI and co-founder of Numeracy Ready. In the past, Emma has been vice principal for teaching & learning, and a good old head of maths.

She found that, while her teachers' subject knowledge was good, "this didn't always lead to good practice in terms of sharing subject knowledge in the classroom." Beyond knowing about the subject, she identified five strands of subject knowledge needed to teach effectively:

1 Exam and curriculum context
2 Progression
3 Multiple methods and representations
4 Misconceptions
5 Probing questions

As part of the course, students prepared several 'Subject Knowledge for Teaching sheets' covering topics of the maths curriculum. The example in Figure 1.6 is by Marcus Bennison. *Marcus is a newly qualified teacher from Brighton and Hove. He started working as a teaching assistant at Portslade Aldridge Community Academy in 2011 before becoming a higher level teaching assistant within the mathematics department in 2013. Marcus studied a BA (Hons) in Secondary Mathematics Education (with QTS) at the University of Brighton. After completing his initial teaching training in June 2017, Marcus started working at Blatchington Mill School, where he now teaches mathematics to secondary school students of varying abilities.* This sheet focuses on addition and subtraction.

Conclusion

Maya was excited by this approach to planning, but she recognised that many of her ideas were not new, especially to expert teachers. Experts draw on mental models of what is to be taught; rather than writing lesson plans, they turn ideas over at odd moments. They have internalised this kind of unit planning and can improvise based on it, drawing on their knowledge of representations and student misconceptions spontaneously (Livingston and Borko, 1989). Maya sees two powerful uses for this kind of planning, however:

1 For teachers new to a course, identifying the knowledge and pedagogical content knowledge needed saves them doing so one lesson at a time, and leaves them far better prepared. Novices plan painstakingly, as they are still creating a mental model of what is to be taught (Livingston and Borko, 1989). Maya believes her approach can help teachers build mental models of what is to be taught more quickly.
2 Teachers and departments can use this approach to codify and collate their experience and wisdom. Lesson plans and slides tend not to transfer between contexts; representations and misconceptions can be used anywhere, tailored to teachers' contexts, needs and preferences. Rather than relying on word of mouth to elicit teachers' experience, Maya hopes that constructing unit plans collaboratively can help teachers share knowledge more effectively.

Having set out to become a more responsive teacher, Maya is surprised to have spent so much time on planning; she is pleased with the clarity she now has about what students

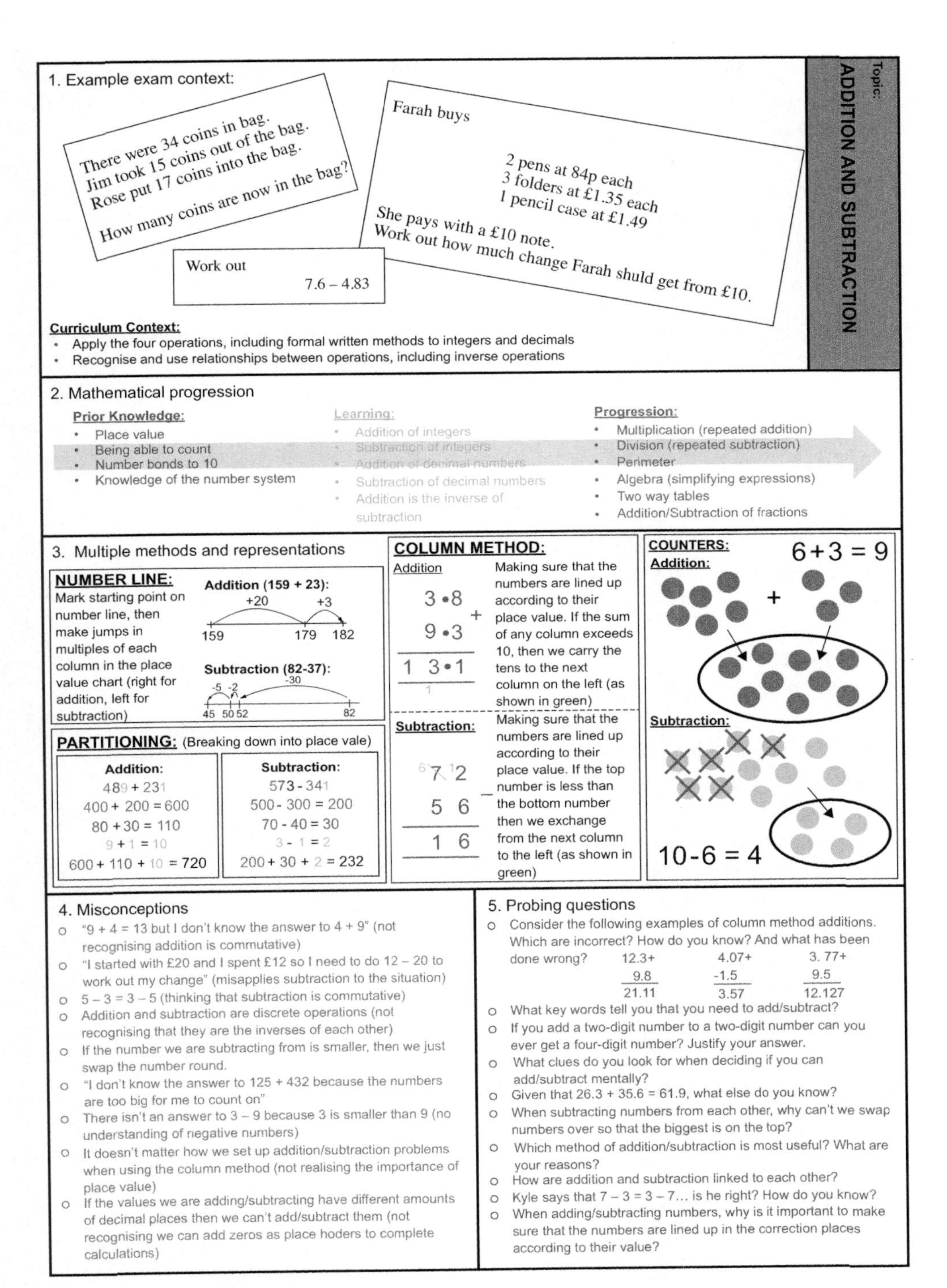

Topic: ADDITION AND SUBTRACTION

1. Example exam context:

There were 34 coins in bag.
Jim took 15 coins out of the bag.
Rose put 17 coins into the bag.

How many coins are now in the bag?

Work out

7.6 – 4.83

Farah buys

2 pens at 84p each
3 folders at £1.35 each
1 pencil case at £1.49

She pays with a £10 note.
Work out how much change Farah shuld get from £10.

Curriculum Context:

- Apply the four operations, including formal written methods to integers and decimals
- Recognise and use relationships between operations, including inverse operations

2. Mathematical progression

Prior Knowledge:

- Place value
- Being able to count
- Number bonds to 10
- Knowledge of the number system

Learning:

- Addition of integers
- Subtraction of integers
- Addition of decimal numbers
- Subtraction of decimal numbers
- Addition is the inverse of subtraction

Progression:

- Multiplication (repeated addition)
- Division (repeated subtraction)
- Perimeter
- Algebra (simplifying expressions)
- Two way tables
- Addition/Subtraction of fractions

3. Multiple methods and representations

NUMBER LINE:
Mark starting point on number line, then make jumps in multiples of each column in the place value chart (right for addition, left for subtraction)

Addition (159 + 23):
+20 +3
159 179 182

Subtraction (82-37):
-5 -2 -30
45 50 52 82

PARTITIONING: (Breaking down into place vale)

Addition:
489 + 231
400 + 200 = 600
80 + 30 = 110
9 + 1 = 10
600 + 110 + 10 = 720

Subtraction:
573 - 341
500 - 300 = 200
70 - 40 = 30
3 - 1 = 2
200 + 30 + 2 = 232

COLUMN METHOD:

Addition

3 •8
9 •3 +
1 3 •1
1

Making sure that the numbers are lined up according to their place value. If the sum of any column exceeds 10, then we carry the tens to the next column on the left (as shown in green)

Subtraction:

6 7 1 2
5 6 –
1 6

Making sure that the numbers are lined up according to their place value. If the top number is less than the bottom number then we exchange from the next column to the left (as shown in green)

COUNTERS:

Addition:

6 + 3 = 9

Subtraction:

10 - 6 = 4

4. Misconceptions

- "9 + 4 = 13 but I don't know the answer to 4 + 9" (not recognising addition is commutative)
- "I started with £20 and I spent £12 so I need to do 12 – 20 to work out my change" (misapplies subtraction to the situation)
- 5 – 3 = 3 – 5 (thinking that subtraction is commutative)
- Addition and subtraction are discrete operations (not recognising that they are the inverses of each other)
- If the number we are subtracting from is smaller, then we just swap the number round.
- "I don't know the answer to 125 + 432 because the numbers are too big for me to count on"
- There isn't an answer to 3 – 9 because 3 is smaller than 9 (no understanding of negative numbers)
- It doesn't matter how we set up addition/subtraction problems when using the column method (not realising the importance of place value)
- If the values we are adding/subtracting have different amounts of decimal places then we can't add/subtract them (not recognising we can add zeros as place hoders to complete calculations)

5. Probing questions

- Consider the following examples of column method additions. Which are incorrect? How do you know? And what has been done wrong?

12.3+	4.07+	3. 77+
9.8	-1.5	9.5
21.11	3.57	12.127

- What key words tell you that you need to add/subtract?
- If you add a two-digit number to a two-digit number can you ever get a four-digit number? Justify your answer.
- What clues do you look for when deciding if you can add/subtract mentally?
- Given that 26.3 + 35.6 = 61.9, what else do you know?
- When subtracting numbers from each other, why can't we swap numbers over so that the biggest is on the top?
- Which method of addition/subtraction is most useful? What are your reasons?
- How are addition and subtraction linked to each other?
- Kyle says that 7 – 3 = 3 – 7… is he right? How do you know?
- When adding/subtracting numbers, why is it important to make sure that the numbers are lined up in the correction places according to their value?

Figure 1.6 Subject knowledge for teaching sheet

Courtesy of Marcus Bennison

are to learn and how she may teach it. She does not see her specification of knowledge and pedagogical content knowledge as a straitjacket; it allows her to be more responsive. As Deborah Ball put it:

> The children's understandings and confusions provided me with more information with which to adapt my choices, yet the mathematics helped me to listen to what they were saying . . . to hear in the students' ideas the overtures to important understandings and insights. (1993, p. 32)

Maya cannot be responsive without purpose and preparation. Her plans now also offer her that purpose and tools with which to respond. Satisfied with her unit planning, Maya turns her attention to planning individual lessons.

Checklist

1 Are you working from a clear curriculum? ☐
2 What will most students already know? ☐
3 What will every student know by the end of the unit?
- Facts ☐
- Terms ☐
- People ☐
- Processes/factors ☐
- Images/diagrams ☐

4 What representations/explanations can convey this knowledge? ☐
5 What misconceptions are likely? ☐
6 How could students make connections to horizon knowledge? ☐
7 How can this be sequenced effectively?
- Revisiting key ideas ☐
- Encouraging connections ☐

Double-check to avoid:

- Vague and abstract statements ☐

A great read on this is . . .

Ball, D., Thames, M. and Phelps, G. (2008). Content knowledge for teaching: What makes it special? *Journal of Teacher Education*, 59(5), pp. 389–407.

Deborah Ball and her colleagues refined Lee Shulman's definition of pedagogical content knowledge, identifying sub-domains including Common Content Knowledge – things which are generally known; Knowledge of Content and Students – how students are likely to respond; and Knowledge of Content and Teaching – effective ways to sequence representations and problems.

2 How can we plan a lesson, when we want students to learn so much, and have so little time?

The problem
There is so much to teach, and not enough time in which to teach it. We cannot respond to student learning until we are sure what matters most.

The evidence
Pick a focus for lessons
Focus on teaching students the subject
Plan the cognitive load students will experience

The principle
Responsive teachers focus lessons on a single, academic purpose.

Practical tools
Pick one focus
Select intrinsic cognitive load, remove extraneous, add germane

Experience Lizzie Strang
The merits and problems of focused objectives

Checklist

? The problem

There is so much to teach, and not enough time in which to teach it. We cannot respond to student learning until we are sure what matters most.

Andy has developed clear unit plans and wants to be more responsive in his teaching. This proves hard: each lesson incorporates a range of knowledge and skill, so which aspects of student understanding, or misunderstanding, should he respond to? Things come to a head with observation feedback from a colleague:

> You said the objective of the lesson was to introduce the class to *A Christmas Carol*. First, you asked students to think of words related to Christmas. Then you asked them

> to write haikus as a group, using some of those words – and since they'd not written haikus since January, you had to reteach haikus. Once they'd shared their haikus you gave them a wordy sheet which outlined the history of Christmas from Pagan times to the present. When I left the room, there'd been no mention of Dickens, you were no nearer the Victorians, and students had learned nothing about *A Christmas Carol*.

Andy wants his students to write haikus, know the history of Christmas and appreciate the context of *A Christmas Carol*; he's appalled at the idea of narrowing his goals. He resists the urge to explain that all knowledge is useful, that he's building his students' teamwork and focusing on responsive teaching; instead, he reviews his lesson planning. Andy realises that before he can share what success looks like (Problem 3), check whether students have achieved it (Problems 4 and 5) and offer useful feedback (Problem 6), he needs to be clearer about what he wants them to achieve. He wants to know:

- What makes a good objective for a lesson?
- What makes a good plan?

The evidence

Pick a focus for lessons
Focus on teaching students the subject
Plan the cognitive load students will experience

Andy does not accept everything he reads, but three bodies of evidence point to a similar conclusion: greater focus will allow him to be more responsive to students' needs.

1. Pick a specific focus

> The best barometer for every lesson plan is 'Of what will it make the students think?'
> (Willingham, 2009)

Andy wonders whether making lessons fun and engaging has proved an unhelpful distraction. He varies tasks to maintain students' interest: making posters one lesson, information hunts around the room the next, social media posts summarising characters' reactions in another. This has been problematic, though. First, introducing novel tasks requires time for explanations and familiarisation. Second, he has been engaging students in the task, not the content; it is the content Andy needs students to value and retain, but he has been selling the lesson by masking the content as something more enticing (Lemov, Driggs and Woolway, 2016). Third, variation may be distracting; thinking about social media posts distracts students from thinking about characters. Memories are tied to the context in which they are learned. Reading a passage describing a house, those asked to read as burglars were more likely to remember that a side door was unlocked; those reading as home-buyers were more likely to remember a leaking roof (Anderson, Pichert and Shirey, 1983). If students' memories are tied to writing social media posts, this will hinder their recall in different contexts.

- **Andy resolves to: "review each lesson plan in terms of what the student is likely to think about"** (Willingham, 2009, p. 79). He will ensure his lessons focus students' thinking on the meaning of the most important ideas. To help them recall the characters in *A Christmas Carol*, he will avoid distracting them with irrelevant tasks.

Andy is happy to focus students' thinking on relevant tasks; he is less happy to read evidence suggesting he should formulate tightly focused objectives. He wants students to write haikus, appreciate Victorian England and understand changing celebrations of Christmas, but the evidence suggests they may need to focus on one goal at a time. Deliberate practice focuses on "well-defined, specific goals . . . improving some aspect of the target performance; it is not aimed at some vague overall improvement" (Ericsson and Pool, 2016, p. 99). Multitasking requires switching between tasks; this "inevitably leads to a loss of concentration, the need for longer periods of study and poorer performance" (De Bruyckere, Kirschner and Hulshof, 2015, p. 97). Andy realises that students would benefit from objectives which allow them to focus on improving at one thing at a time. Andy has always followed the advice of Dylan Wiliam (2011) to consider the context of the lesson and the objective separately. For example:

- **Objective:** writing a letter.
- **Context:** expressing Romeo's feelings on having met Juliet.

Andy hoped that students would improve at letter writing and expressing Romeo's feelings; he wonders now whether letter writing distracted students from understanding *Romeo and Juliet* better. He is concerned that objectives like this imply that writing a letter is a transferable skill, when a good letter demands knowledge of the context (see Introduction). Students need to know how to do a task well (see Problem 3), but sharing models of 'double-content' tasks (writing a letter *and* expressing Romeo's feelings) seems to be no better than teaching them separately, and may even detract from learning (Renkl, Hilbert and Schworm, 2009). He decides that 'explaining Romeo's feelings on having met Juliet' would be a perfectly good objective; mixing this with letter-writing limits students' ability to succeed in either objective. Andy concludes that he needs to focus on a single objective at a time, and consider the context as the objective.

- **Andy resolves to focus on a single objective, and its context, at once:** he will allow students to focus on tasks designed solely to promote their understanding of *A Christmas Carol*.

2. *Focus on teaching the subject*

Andy has always hoped to build students' confidence, motivation and self-efficacy during lessons, but he reconsiders how they are developed as he reads the evidence. He knows that skills such as creativity, collaboration and critical thinking rely on students knowing the subject (Introduction); he is surprised to learn that students' motivation and confidence may rely on learning the subject, too.

> Teachers who are confronted with the poor motivation and confidence of low attaining students may interpret this as the cause of their low attainment and assume that it is both necessary and possible to address their motivation before attempting to teach them new material. In fact, the evidence shows that attempts to enhance motivation in this way are unlikely to achieve that end. Even if they do, the impact on subsequent learning is close to zero (Gorard, See & Davies, 2012). In fact the poor motivation of low attainers is a logical response to repeated failure. Start getting them to succeed and their motivation and confidence should increase.
>
> (Coe et al., 2014, p. 23)

Self-efficacy - confidence in one's own abilities - is domain-specific: a student may feel brilliant at geography and hopeless in Spanish. The biggest contributor to self-efficacy is experiences of mastery; the second-biggest is seeing others succeed (Bandura, 1982). Andy can instil confidence by supporting his students to achieve academically and to recognise and reflect on their growth. This makes more sense than trying to instil unwarranted confidence - which was always a struggle anyway, since students are not easily fooled. Andy needs to instil enough confidence and motivation to ensure students put pen to paper, but once they have started, their confidence and motivation will grow as he shows them what success looks like (Problem 3) and helps them to achieve it.

Andy wants to develop students' wisdom and maturity, too, providing opportunities for them to gain flashes of insight about themselves, one another and the world. He is concerned to learn of the lack of reliable measures and definitions of character development, let alone widespread evidence that it can be done effectively (Didau and Rose, 2016). He struggles to conceive of activities which help students develop wisdom and maturity while also teaching quadratic equations; whenever Andy has tried, he has focused on social goals, squeezing out academic learning, or he has added cosmetic features to tasks which detract from learning. Mixing purposes has diminished both. Character, creativity and cognitive flexibility seem to develop from experiencing disorienting experiences like foreign travel and violations of the laws of physics (Ritter et al., 2012; Lu et al., 2017). Andy is inspired by the National Trust's list of fifty things to do before you're 11¾ - go stargazing, roll down a hill, find your way with a map and compass (National Trust, n.d.). He realises, with regret, that these are things for co-curricular activities and days off timetable, not lessons; it is hard to replicate the experience of rolling down a hill between 10.30 and 11.30 on Tuesdays while teaching the past tense. Andy wants to help his students gain maturity and wisdom, but he resolves to prioritise teaching the knowledge they need to understand and contribute to the world during lessons, dedicating himself to broader educational goals as a tutor and in leading co-curricular activities.

- **Andy focuses on the subject during lessons and on students' maturity and wisdom outside lessons.**

3. *Plan for cognitive load*

Cognitive load theory explains much of what Andy has discovered about the need to focus on the subject and suggests ways he can do so better. The theory focuses on learning as the creation of schemas: organised structures of knowledge in long-term memory. Ideas reach long-term memory having been held in working memory: what we are conscious of at a given moment. Working memory can only cope with a limited cognitive load; people can retain a handful of isolated facts (a few numbers, for example) or process two or three ideas (Sweller, van Merriënboer and Paas, 1998; Cowan, 1999). Therefore, the load placed upon working memory - what students are thinking about - is crucial for learning. An early study divided students into two groups that approached the same trigonometry problems with different goals:

- Group A were given a specific goal (find the length of a particular line).
- Group B were not given a specific goal (find the length of any lines you can).

They were solving identical problems, and Andy's first guess was that Group B would do better, since they had clearer instructions. In fact, students in Group A endured greater cognitive load, because they had to keep in mind numerous ideas which did not help them think about the trigonometry itself: what they were trying to find, what they had already found and what they needed to find next. Students in Group B remembered more about the problems and learned more, whereas for students in Group A, "the cognitive-processing capacity needed to handle this information may be of such a magnitude as to leave little for schema acquisition, even if the problem is solved" (Sweller, 1988, p. 261). Students can think about solving problems and they can think about tasks which contribute to their long-term memory, but if the task is challenging for them, their working memory is unlikely to be sufficient both to solve the problem and to recall key ideas about the task. Working memory limitations restrict how much of an experience reaches long-term memory; designing teaching "which flouts or merely ignores working memory limitations inevitably is deficient" (Sweller, van Merriënboer and Paas, 1998, p. 253). Andy believes that planning using cognitive load theory should help students learn more.

Andy finds the three forms of cognitive load proposed by Sweller, van Merriënboer and Paas (1998) offer a helpful frame for planning:

1 **Intrinsic cognitive load:** the challenge of learning complicated ideas. Individual facts can be learned in isolation, so have low intrinsic cognitive load. Students can learn the meaning of 'Je' without knowing the word 'aime' in French; they can learn that Fe is iron without having to remember that Cu is Copper. Some learning has high intrinsic cognitive load, however. Students must consider the relationship between each word in a sentence to ensure it is grammatically correct; explaining a chemical reaction requires understanding the interaction between different elements. Whether learning has high intrinsic cognitive load reflects a student's existing knowledge, not an absolute standard; a competent French speaker may face low intrinsic cognitive load when considering the

words in a simple sentence, but need to think more carefully when using the subjunctive tense. It is possible to break down content with high intrinsic cognitive load (van Merriënboer and Sweller, 2005), but Andy's priority is selecting the intrinsic cognitive load on which the lesson will focus carefully.

2 **Extraneous cognitive load:** the distraction caused by tasks which occupy working memory but do not contribute to the formation of long-term memories. This includes:
   - Splitting attention: asking students to refer to two sources of information simultaneously. Andy can avoid this by ensuring information, such as labels, is where it will be needed.
   - Redundancy: information which adds nothing detracts from learning. Andy can avoid this by removing irrelevant and distracting labels and text.
   - Expertise reversal: support which helps novices, like model answers, can hinder experts. Andy monitors students' success closely and removes support as students become more skilled.

   Andy may use the modality effect, too: describing an image verbally, for example, talking through a diagram, effectively increases students' working memory.

3 **Germane cognitive load:** additional cognitive load which contributes to the formation of long-term memories. It can be created by tasks which depress student performance initially but help retention, including spacing learning over time (Pashler et al., 2007) and varying practice (for example, varying the types of mathematical questions students are answering in a lesson; (Soderstrom and Bjork, 2015)). This is not because errors are desirable in themselves, but because some changes, like varying practice, may cause more errors initially, but better retention.

Andy recognises limits to what cognitive load theory offers. The research findings have been tested in a handful of topics, subjects and lessons. While they seem robust, and work to extend them is ongoing, only a few examples have been tested in classrooms, so he needs to apply the ideas carefully. Varying practice will work in a grammar lesson, but in a lesson on literature it may prove unhelpfully confusing; instead, Andy may simply add a retrieval question to his exit ticket (Problem 4) to space learning (Pashler et al., 2007). He is also aware that approaches which support learning, like varying practice, can prove disorienting and hence unpopular with students (Brown, Roediger and McDaniel, 2014, p. 54). He will explain the rationale for his choices to students and perhaps even occasionally choose less efficient ways of teaching to maintain student enthusiasm. Andy finds that cognitive load theory offers useful ideas, not perfect answers; applying it requires tailoring the principles to his students and lessons.

- **Andy plans for cognitive load by:**
  - **Choosing intrinsic cognitive load carefully.**
  - **Cutting extraneous cognitive load: removing unproductive distractions.**
  - **Increasing germane cognitive load judiciously: adding challenges to increase recall.**
  - **Recognising the limits of the existing research, as well as its promise.**

The principle

Responsive teachers focus lessons on a single, academic purpose.

Practical tools

Pick one focus
Select intrinsic cognitive load, remove extraneous, add germane

Andy begins with his knowledge organiser and unit planner (Problem 1). Based on what he plans to teach and what students learned in the previous lesson (Problem 4), he:

- Decides a single focus for the lesson and then identifies:
    - What is critical to that focus?
    - What supports that focus?
    - What is optional?

- Ensures every task focuses students' thinking and schema construction on the objectives by:
    - Identifying what tasks to focus students' thinking upon
    - Minimising extraneous cognitive load
    - Introducing germane cognitive load

Version 1: planning from activities

Andy reflects guiltily on his experience planning from activities. With an hour to fill and a quick glance at the textbook, he would identify the main point of the lesson: the food chain, for example. He would plan a hook to engage students: 'A rabbit lives in the forest. What should it be afraid of? What will it live on?' He would then choose something to get students working on the key ideas, such as placing chunks of information around the room for students to find and record. Students would order the information to show which organisms consume which and finally create a poster showing a full food chain. Andy realises he was primarily fitting activities together to fill the lessons (in common with 85% of teachers, according to one survey; Fryer, 2017). This did not guarantee much student learning. Andy was aware of the key points and misconceptions, but whether they emerged depended on what occurred to him and the questions students asked. Planning based on activities did not create the focused lessons Andy wanted: he needed to plan backwards, from his objectives.

Version 2: planning from objectives

Andy formulates objectives, deciding he wants students to:

- Remember the meaning of producers, consumers, predators and prey.
- Locate organisms on a food web.
- Explain energy transfer within a food web.

- Analyse interdependence in a food web.
- Apply these ideas to a new ecosystem.
- Evaluate the dangers of toxic material in the food web.

These objectives are clearer, but they do not allow Andy to plan in a focused way. The objectives justify – or describe – activities, rather than the underlying objectives: 'locate organisms on a food web' will help students explain energy transfer within food webs; 'use similes in a range of sentences' will help students write an elegant description of a landscape. Additionally, Andy does not plan to teach some of the things he has listed as objectives. Students have learned how to place organisms on a food web already; Andy just wants to check their existing knowledge. Andy has written objectives for the activities he plans to use, but this leaves his underlying objectives unclear. If something is an objective, Andy needs to offer his students models, practice, feedback and time; he can only do so by choosing fewer, more focused objectives.

Andy has tried to ensure the objectives become more difficult, but this begins to seem confused to him. He has used different verbs from Bloom's taxonomy, so students move to progressively more challenging objectives, from 'remembering' to 'evaluating'. Andy worries that this may encourage students to underestimate the importance of remembering key ideas. He realises he could move the verbs around very easily: describing something accurately can be as hard as analysing it; asking students to 'explain interdependence in a new ecosystem' would be just as hard as asking them to 'apply these ideas to a new ecosystem'. The real difficulty in lessons comes from the intrinsic cognitive load. Placing organisms on a food web is relatively simple; interdependence is more challenging, because it requires students to consider the interaction of many different organisms. Relying on Bloom's taxonomy to formulate challenging objectives seems ineffective: Andy decides to rewrite his objectives more carefully.

Version 3: specifying key objectives

Andy revises his objectives to distinguish between what he wants students to improve upon and other activities which support this. He identifies two critical objectives:

1. Explain the transfer of energy within food webs.
2. Explain interdependence in food webs.

Andy recalls the problems with vague, decontextualised objectives. He does not expect all of his students to be able to explain interdependence in any food web at the end of this lesson, only the woodland food web they will be studying. He revises the objective to specify:

1. Explain the transfer of energy within a woodland food web.
2. Explain interdependence within a woodland food web.

These are the critical objectives for the lesson. They are hard; Andy does not need to use words like 'evaluate' or 'analyse' to make interdependence challenging.

Two of the objectives Andy had set are not actually goals for improvement during the lesson:

1 Know the meaning of producers, consumers, predators and prey.
2 Place organisms on a food web.

They are prerequisites for Andy's objectives - students will not be able to explain energy transfer if they do not know the difference between a producer and a consumer. Andy hopes students can do this already, so they are not learning objectives, but he will check students' knowledge and remind them if necessary.

Andy chooses two further objectives which are desirable, but not critical:

1 Apply interdependence and energy transfer to a marine ecosystem.
2 Explain the dangers of toxic material entering the marine food web.

This is more challenging, asking students to transfer their understanding of a woodland ecosystem to a marine one and drawing on horizon knowledge (Problem 1). These objectives also have higher intrinsic cognitive load. For students to meet the final objective, they have to understand both energy transfer and the effect of toxic materials. Andy hopes students will get this far, but it is not essential this lesson and he can address these objectives next lesson if necessary.

Andy is happy with his objectives. He has a clear focus: two main objectives which are specific and manageable and more challenging objectives with greater intrinsic cognitive load. He has identified the prerequisite knowledge for students and can plan accordingly. He has avoided relying on increasingly complicated verbs to make his objectives challenging. Andy realises intrinsic cognitive load could help his colleagues increase the challenge in their lessons, too. In a history lesson, students might begin by learning individual problems Henry VIII faced, and go on to compare the importance of these problems. In maths, students who know that 3^2 is 3×3 might be challenged to answer $4 + 3^2$. In each case, the first is a prerequisite for the second, and fine judgement is required not to overload students. Intrinsic cognitive load seems to provide a useful, usable framework.

Having formulated clear objectives, Andy designs tasks and activities which will achieve them, but he is dissatisfied with the results. He returns to his information hunt and changes the facts slightly to go beyond creating a food web and include energy transfer. He will attach twenty facts to classroom walls, such as 'barn owls eat three to four rodents a day'; students are asked to collect them as fast as possible, then create a poster which shows energy transfer and interdependence, which they will present to the rest of the class. Andy realises this lesson plan subjects students to extraneous cognitive load: additional tasks which distract from adding to their long-term memories. Copying information not only risks students making mistakes, but also splits their attention and distracts them from the meaning of the information. Creating a poster summarising this information has students thinking about how to show a food chain, not the energy transfer within it. Finally, while presenting findings offers practice communicating, limited time for preparation and shallow understanding mean students may learn little while presenting or listening. Andy's students will be distracted by extraneous cognitive load; his plan confuses disentangling and presenting information with thinking to form long-term memories about key ideas.

Version 4: removing extraneous cognitive load

Andy removes extraneous cognitive load. He will model how students can explain energy transfer and interdependence (see Problem 3), showing images demonstrating how energy is transferred from producers to consumers and then asking students to represent energy transfer by completing pyramids of biomass. Andy removes redundant tasks, like students drawing their own food webs; this focuses their thinking away from the key ideas of energy transfer and interdependence. He removes distracting information, too, like unhelpful labels, to help students focus on the ideas he wants them to remember most. Happy that he has removed extraneous cognitive load from the lesson, Andy wonders whether he can add germane cognitive load.

Version 5: adding germane cognitive load

Andy identifies additional changes which may challenge students initially, but will increase their chances of remembering key ideas. He decides to check prerequisite knowledge of vocabulary and food webs using retrieval practice; a quiz will tell him whether students have remembered what they need to know for the lesson and increase the chances they will remember next time. Just as his sports coach varies the practice in Andy's hockey games, he plans to vary students' practice in the hope it will get them closer to understanding the underlying ideas, not just the surface features. Once students have explained energy transfer within a woodland food web, he will ask them to do the same for a marine ecosystem, and perhaps even a desert. Similarly, once he has taught students interdependence in woodland food webs, he will ask them to apply the same ideas in marine and tundra ecosystems. While Andy expects students will perform worse during the lesson, they should remember more when he returns to this next lesson - he makes a mental note to mention this to students early on to avoid their becoming discouraged.

Reintroducing variation

As planning for cognitive load becomes habitual, Andy worries that his lessons are becoming tedious. They are predictable, lacking the buzzy activities students and visitors seem to appreciate, like group activities and role play. Partly, Andy feels assured that students are learning more and are challenged more productively, but he reintroduces more imaginative activities occasionally, confident that attention to objectives and cognitive load will ensure they are purposeful. He uses his collection of representations (Problem 1) to present ideas in interesting ways. With clear objectives and a good exit ticket (Problem 4), Andy is confident he will identify any gaps in students' knowledge and will not be misled into believing that students' enjoyment of a task guarantees they are learning.

Andy realises that creative activities are more powerful once students know a topic. Visiting a history lesson, Andy is envious to see a brilliant debate between key figures in the struggle for Indian independence. His colleague explains, however, that it relies on the preceding lesson; students read, summarised and discussed each figure's view, and had a strong grasp of the lesson content before it began. Andy has been frustrated when student responses to creative tasks rely on their skill in art, theatre or music, or just their prior knowledge. Now that students are learning more, they can apply their knowledge to creative tasks better; provided with clear instructions which focus their thinking on the key ideas, students can play the characters in a lab, a play or a historical era effectively.

Andy also wonders whether he is teaching spontaneously enough. Before he had clear objectives, he diverted the lesson readily in response to student questions. Now that he has established how much he wants students to learn (Problem 1) and has formulated clear objectives, he is more reluctant to do so. Andy concludes that tight objectives and carefully chosen activities do not preclude adaptation, and he is happy to divert the lesson when:

- Diversion is essential to achieving the lesson's objectives.
    - If students had never seen a food web before, contrary to his expectations, he would have dedicated time to teaching this.
- He can link the diversion to the lesson's objectives.
    - A question about the significance of the dogs in *Animal Farm* during a lesson on the parallels with the Russian Revolution provides another example for students to consider, even if Andy did not intend to discuss the dogs in this lesson.
- Diversions help students make connections to past topics.
    - When students notice similarities between Charles I's Personal Rule and Henry VII's consolidation of power, Andy asks them to explain the similarities and differences, strengthening their understanding of both.
- Diversions introduce future topics.
    - When students confuse the method for obtaining a fraction with that for obtaining a ratio, Andy explains that this is how a ratio is obtained, preparing students for future lessons.

With a clear sense of his goals for units and lessons, Andy can connect students' questions to the most important ideas and resist the temptation to take an intriguing diversion which will prevent students from achieving the day's objectives. When Andy can see a productive destination for a diversion and how he will help students make that link, he takes it; where he cannot, he avoids the diversion.

Andy decides that sometimes academic objectives are less important than social ones. In the first lesson of the year, Andy signals the importance of learning by introducing key ideas immediately, but with a group that has not been taught together before, he reserves time for students to get to know one another. After an offensive comment from a student, he spends time establishing exactly why the comment is inexcusable. While cognitive science suggests tasks which may unsettle students, like constantly varying practice, where students lack confidence, he designs something less efficient as a step towards the more efficient method. Finally, Andy feels that students need a break occasionally, and he lets them do something less taxing – as when they come to his lesson between one exam in the morning and another in the afternoon. Learning and social goals often pull Andy's planning in different directions. Rather than convincing himself he can do both simultaneously, he decides to pursue one or the other wholeheartedly, aware of the value it has and reconciled to what is lost as a result.

Experience – Lizzie Strang

The merits and problems of focused objectives

Lizzie Strang has been a lead practitioner and English teacher, and now teaches history at Camden School for Girls in London. She is currently completing an MSc in Learning & Teaching at Oxford University.

Why did you come to make objectives narrower and more specific?

I think this was a combination of wanting my students to get better at text analysis, which led me to having a greater appreciation of the centrality of knowledge in skill building. As a result, I started thinking more specifically about not only what I wanted the students to do but also to know by the end of the lesson. This was then reflected in my lesson objectives.

For example, a typical learning objective earlier in my career on Shakespeare's *Macbeth* might have been: 'To identify features of Shakespeare's use of language and explain the effect on the reader'.

An issue with an objective such as this is that its vagueness means the lesson lacks direction and therefore rigour. Not establishing exactly what I want students to identify and explain the effect of meant that I couldn't prepare ways to stretch their thinking about it and as a result be better placed to hone their skills of analysis.

I know when I read *Macbeth* there are particularly important ideas, themes, language choices/dramatic techniques that are more valuable than others. With a learning objective like this, I leave it to chance that students will pick up these ideas.

The learning objective also suggests that what is valuable in this lesson is spotting language devices and explaining what they do. This isn't what the lesson is about at all! No, this lesson is about engaging with *Macbeth*'s growing sense of guilt and appreciating how Shakespeare makes this come alive to us through the images he creates with words. Disconnecting the content from the skills we want to develop leaves us with an aim that is hollow.

What did you change?

Today a similar objective would be: 'How does Shakespeare suggest that *Macbeth* is disturbed by his guilty conscience in A2S2?'

In order to resolve the issues above I needed to rethink my planning so this is what really changed. After setting out a lesson objective I am now more prescriptive in what I want students to know and do by the end of the lesson, often writing something akin to a 'gold standard' response. For example, I might write a note to myself something like:

> [students to comment on how the stage direction and line – 'looking at his hands this is a sorry sight' suggests Macbeth instantly regrets murdering Duncan . . .]

It doesn't have to be a full-on paragraph but I do find writing out the kind of response (and therefore the kind of language/thinking) I'm looking for as opposed to having an outcome like 'students explain the effect of the stage direction' helps me teach more purposefully, with more rigour and higher expectations.

What barriers and tensions have you found between specifying outcomes and wanting individual responses? How did you address them?

As a teacher of English, I always want to work towards enabling my students to find their own voice and their own interpretation, and to appreciate that there could be multiple valid interpretations of a text. The reason why I teach English is a love of the open-ended nature of the subjects. I feel therefore there is a tension between teaching English in a way that feels as though it honours the discipline's 'openness' and teaching it in a way that is more instructive, even prescriptive!

In terms of addressing this tension, I'm still grappling with it! In one sense, I think it is a healthy, natural tension between the subject as a discipline and the subject as part of the school curriculum. I know I am doing my students a service by being much clearer in my own mind about what exactly I want my students to know and do by the end of the lesson. Across the course of the term and year I will of course work towards giving students more freedom in their interpretations, built on a foundation of guided practice and a detailed knowledge of the text.

Overall tips

I think because we use the term 'lesson objective' so often (usually prescribed as part of a scheme of work) the term has become slightly disconnected with its meaning and the actual process of lesson planning.

A lesson objective should reflect the knowledge and skills you want the students to have thought about and practised by the end of today's lesson. Crystallise what you are looking for by clarifying what a 'gold standard' answer would be, write this down and plan backwards from there.

If you're just beginning as a teacher and not sure exactly what you want your students to do and know by the end of the lesson; it's what you think are the important take-aways about that scene, chapter, topic etc. . . .

Conclusion

As a new teacher, Andy tried to make every lesson different and every moment exciting. In retrospect, designing and explaining novel activities sapped planning and lesson time and imposed heavy extraneous cognitive load as students wrestled with activity instructions rather than learning. Camouflaging the learning with fun activities did not interest students in the subject; establishing specific, subject-focused objectives makes lessons simpler to plan and more productive for students. Andy realises now where his English lesson went astray: many of his tasks did not add to students' schemas for *A Christmas Carol*. Planning this again, he would identify exactly what he wanted students to know. Cognitive load theory and specific objectives do not solve every aspect of planning, but they provide a powerful way to think about learning.

Andy seeks to be an expert teacher, "one who chooses to use a particular teaching procedure at a particular time for a particular reason" (Loughran, Berry and Mulhall, 2012, p. 2). He is content provided he has a clear goal, knows of what an activity will make students think and is conscious of the trade-offs he is making. Andy realises how powerful it is to be specific and feels that he is now ready to take on the challenge of responsive teaching properly. Clear about what he wants students to learn, he can share what students need to do (Problem 3), identify whether they have achieved it after (Problem 4) or during (Problem 5) lessons and provide useful feedback (Problem 6).

Checklist

1. What should students know/be able to do by the end of the lesson?
   - What is critical? ☐
   - What supports that? ☐
   - What is optional? ☐
2. Of what will activities make student think? ☐
3. How do the objectives apply increasing intrinsic cognitive load? ☐
4. Where is the extraneous cognitive load? How can it be removed? ☐
5. Where could you add germane cognitive load? ☐
6. How can you manage students' responses to challenge? ☐
7. How and when may it be appropriate to divert from your plan? ☐

Double-check to avoid:

- Vague objectives which do not specify context ☐
- Multiple focuses ☐
- Extraneous cognitive load ☐

A great read on this is . . .

Sweller, J., van Merriënboer, J. J. and Paas, F. G. (1998). Cognitive architecture and instructional design. *Educational Psychology Review*, 10, pp. 251–296.

John Sweller and colleagues review Cognitive Load Theory, defining three kinds of cognitive load: intrinsic, complicated material; extraneous, poorly-designed instruction; and germane, complication which depresses performance but increases learning. The authors offer a range of techniques which put these principles to work, like varying practice, avoiding splitting attention and using dual coding (linking words and images).

3 How can we show students what success looks like?

The problem
Students struggle to identify what success looks like.

The evidence
Sharing a sense of how to create success is challenging; examples are critical
Demonstrating what success looks like helps all students, especially low attainers
Knowing what success looks like promotes metacognition and motivation

The principle
Responsive teachers show students what success looks like

Practical tools
Share model work of different standards
Help students identify what makes work great

Experience – Michael Pershan
Helping students see what they can't yet see

Checklist

? The problem

Students struggle to identify what success looks like.

Anne has a clear purpose and objective, but she is struggling to help her students write coherently. They write willingly, using the language structures Anne teaches, but Anne finds their work is leaden where it should be light; imitative where it should be original. Today, Emma's piece is at the top of the pile of marking. Anne steals a glance, then places it carefully at the bottom: Emma *gets* good writing. She takes the structures Anne teaches and transforms them into original pieces which really work; today, Anne asked students to create a feeling of dread; Emma's piece makes Anne shiver. Can this be taught, or is Emma just a born writer? Anne wonders:

- How can she show students what success looks like, and how can it be achieved?

The evidence

Sharing a sense of how to create success is challenging; examples are critical
Demonstrating what success looks like helps all students, especially low attainers
Knowing what success looks like promotes metacognition and motivation

Anne's students must know what success looks like if they are to achieve it. Clear, challenging goals improve performance, directing attention, increasing persistence and arousing relevant knowledge; they increase people's belief they can do well and their motivation to conduct a task (Locke and Latham, 2002). Yet distant goals do not mobilise effort, direct action or support gauging self-efficacy; big tasks need to be broken into "attainable subgoals" which provide "immediate incentives and guides for action" (Bandura, 1982, p. 134; Bandura and Schunk, 1981). Anne needs to share clear, immediate goals: a vision of success. This is echoed by research in deliberate practice, which suggests that it

> both produces and depends on effective mental representations. Mental representations make it possible to monitor how one is doing, both in practice and in actual performance. They show the right way to do something and allow one to notice when doing something wrong and to correct it.
>
> (Ericsson and Pool, 2016, pp. 99–100)

Conveying what success looks like to create these mental representations is challenging, however. Teachers often struggle to describe

> exactly what they are looking (or hoping for), although they may have little difficulty in recognizing a fine performance when they see one . . . Teachers' conceptions of quality are typically held, largely in unarticulated form, inside their heads as tacit knowledge.
>
> (Sadler, 1989, p. 126)

Concrete descriptors, like 'include quotations', 'label the axes' and 'use arms for balance' can help students follow procedures. They cannot convey the quality of what is to be achieved, however. 'Support your argument', 'be ambitious' and 'approach the task methodically' say little; if students knew how to be 'methodical', they would. So, such descriptors can "lead to frustration because of their inflexibility" (Sadler, 1989, p. 134); students' work may meet the criteria but still fail to reach the desired standard. Concrete descriptors do not reflect underlying, implicit strengths of student work (Hammond, 2014). They cannot articulate the tacit (Rust, Price and O'Donovan, 2003): an idea "which cannot be specified in detail cannot be transmitted by prescription, since no prescription for it exists. It can be passed on only by example from master to apprentice" (Polanyi, 1962, p. 53). Students' sense for success is "'caught' through experience" (Sadler, 1989, p. 135): the acquisition of an "arsenal of exemplars" (Kuhn, 2001, cited in Christodoulou, 2017, p. 96). Ideally, these exemplars should be examined by the student alongside an expert. Historically this was done by individual tutors, yet Anne must find a way to share a sense of success with her entire class (Sadler, 1989; Rust, Price and

O'Donovan, 2003). Anne cannot tell students what success looks like; she needs to find a way to show them.

- **Anne realises that a list of criteria helps students check aspects of their work, but does not convey the sense of quality she wants them to gain; this requires sharing model work.**

Anne finds that research in cognitive science has identified two approaches to help students recognise and achieve success.

1) The worked example effect

Studying models – worked examples – has a powerful impact on the quality of student work (Kalyuga and Sweller, 2004; Sweller, van Merriënboer and Paas, 1998; Wittwer and Renkl, 2010). Students who have seen model work seem to learn better and more efficiently (Zhu and Simon, 1987), an effect which has been demonstrated for tasks with clear solutions (like maths problems) and those with a range of solutions (like English essays; (Kyun, Kalyuga and Sweller, 2013)), and is being extended to behaviours like cooperation and subjects like art (Renkl, Hilbert and Schworm, 2009). Successful learning from worked examples "does not always occur naturally however" (Wittwer and Renkl, 2010, p. 394); students may not engage with them (Sweller, van Merriënboer and Paas, 1998).

2) The completion problem effect

A powerful variation of the worked example effect engages students in what success looks like by offering partially completed models and asking students to complete the missing steps. Checking whether students know the next step also provides a rapid test of prior knowledge (Kalyuga and Sweller, 2004). This seems to decrease extraneous cognitive load, helping students to create mental models and transfer their learning to new problems (Sweller, van Merriënboer and Paas, 1998).

Anne looks to introduce worked examples and completion problems within teaching sequences. 'Interleaving' seems particularly effective, alternating between showing worked examples and asking students to solve problems (Pashler et al., 2007). Teachers' explanations seem to add little or nothing to students' understanding of examples (underscoring the challenge of conveying success in words); having students explain the merits of a model themselves may be sufficient (Renkl, Hilbert and Schworm, 2009; Wittwer and Renkl, 2010). What matters is students' thoughtful engagement with the models.

- **Anne resolves to:**
  - **Use worked examples**
  - **Use completion problems**
  - **Interleave models with problems for students to solve**

Clarifying what success looks like seems to benefit all students, but particularly those with low prior attainment. Students with low prior knowledge of a topic benefit from worked examples; those with high prior knowledge can solve problems with limited guidance (Kalyuga and Sweller, 2004; Sweller et al., 2003). Students with low prior attainment struggle with ambiguous goals. For example, higher-attaining students in maths recognise that 1+2 is a procedure (1+2) and a concept (all the ways in which 3 can be reached); lower-attaining students see it as a procedure alone. In consequence,

> the more able recognise that the process of addition first taught is not the main aim of the game and go on to succeed. The less able who try to do what they are asked: to master the counting procedure, seem cheated because when they finally do so, the game has moved on to a more advanced stage and left them behind.
>
> (Gray and Tall, 1994, p. 19)

Overcoming ambiguity by showing what success looks like seems to particularly benefit lower-attaining students. Students who were shown how their work would be assessed, and evaluated their and their peers' work, gained more factual and conceptual knowledge and greater skill in inquiry design than students with the same teacher who were not shown how their work would be assessed. Students with low prior attainment benefited most; the gains "actually enhance" their learning "without impeding the high achieving students" (White and Frederiksen, p. 91). Anne realises that sharing what success looks like has great promise for the students who most need her help.

Additionally, Anne learns that sharing what success looks like seems to support student metacognition and increase motivation. Metacognition - students monitoring their learning and adapting accordingly - seems to have powerful effects on learning, particularly for lower-achieving and older students (Casselman and Atwood, 2017; Koriat, 2007; Ericsson and Pool, 2016; Zimmerman, 2002). Student self-monitoring relies on accurate self-assessment: the student coming "to know what constitutes quality" (Sadler, 1989, p. 126) and how their work compares to it. Helping students to self-assess accurately and plan accordingly has powerful effects on learning (Casselman and Atwood, 2017; White and Frederiksen, 1998). Knowing what success looks like is also motivating. White and Frederiksen (1998) found that students with low prior attainment were prone to submitting duplicate projects (copying their partners' work), but in the group shown what success looked like, this "tendency was nearly eliminated" (p. 37). This seems to be a strong basis for independent learning and collaboration; when Anne sees Sixth Form students collaborating, she realises this is because several years of learning the subject has shown them what success looks like, thus allowing effective collaboration.

- **Anne hopes that sharing what success looks like will encourage students and improve their metacognition.**

The principle

Responsive teachers show students what success looks like

Practical tools

Share model work of different standards
Help students identify what makes work great

Anne wants to:

- Share models which show students how to succeed in their tasks.
- Support students to identify what about those models matters.
- Help students use these examples to improve their own work.

Version 1

Anne has formulated clear objectives for the lesson (Problem 2), and she tries using these to share success criteria with students. She considers involving students by having them suggest learning objectives, rewrite objectives in their own words or guess missing words from the objectives. Anne notes a significant limitation, however: students might state the goal ('Write more elegantly') or they might state how it will be achieved ('Choose more specific verbs'), but this offers no guarantee they will be able to pick better verbs while writing. So Anne tells students the lesson's objective, sometimes through direct statements ('This lesson we're going to learn to multiply fractions'), sometimes by showing a problem students will be able to complete by the end and sometimes by providing introductory stimulus material such as an unexpected image and helping students frame worthwhile questions (Phillips, 2001). Sometimes, Anne asks students to recall the objectives and link them to the current activity. Sharing the objectives seems motivating and worthwhile; she wants students to know the goal of each lesson. She recognises, however, that this does nothing to show what success looks like; she needs to introduce students to model work.

Version 2

Lessons and units build towards work which encapsulates the learning: a good sentence, solution or essay; Anne wants students to know how to complete them well. She considers sharing the mark scheme with students, going through the descriptors together or rewriting them collectively to clarify what she's looking for. She begins writing success criteria:

- Use rhetorical questions.
- Give your opinions.

Anne's experience is that lists like this encourage students to use rhetorical questions frequently and carelessly, however. Lists are helpful for improving some aspects of students' work - checking decimal points and capital letters, for example - but she wants students to understand when a rhetorical question will work. Anne has been trying to describe good writing, to convey tacit ideas in words, and has been failing. Offering further description when her first description did not work seems unhelpful, while providing increasingly specific criteria about how students are to respond will constrain their writing severely. Anne notes that mark schemes often lack ambition and are unhelpful in showing what matters (Hammond, 2014; Massey, 2016). Only one success criterion counts: is it good? For persuasive writing, for example: 'Did your writing prove persuasive?' In maths: 'Is this a clear, accurate solution?' The challenge is sharing the ingredients of persuasion and clarity with students without resorting to a focus on mechanical features. Raking over descriptions of success seems less helpful than examining examples.

Version 3

Anne is inspired by Ron Berger, who describes how examples "set the standards for what I and my students aspire to achieve" (2003, pp. 29-30). Berger offers students "a taste of excellence" (p. 31) using models by former students and videos of them presenting their work, alongside models from other schools and from the professional world. Carolyn Massey demonstrates historical writing by examining the work of historians such as Orlando Figes (2016). Anne identifies an excellent answer to share with students. She is concerned that if she leaves students with the model answer they may copy it, but decides that imitation is part of apprenticeship; removing models prevents students from returning to them as they work. She is also confident that any copying will be obvious: she decides to leave students with the model work. She wonders whether showing students one model will be sufficient, however; students may not be able to tell the difference between a good piece of work and a mediocre one. She needs to help them compare.

Version 4

Anne decides to show models of differing levels of quality. Contrasting examples seems to help students understand the criteria for great writing, improving their self-assessment, their choice of strategies to improve and their writing (Lin-Siegler, Shaenfield and Elder, 2015). Comparison seems

> to have helped students generate the differentiated knowledge structures that enabled them to understand deeply what specific good features to include and what specific poor features to avoid in the story writing. This resulted in recognition of poor aspects of writing, generation of strategies to correct them, and willingness to revise and make improvements. (p. 531)[1]

Anne identifies a model of excellent work from the previous year, borrows an average model from two years ago from a colleague and writes a poor model which incorporates many common errors herself. Anne decides to share the models early in the unit and to ask students to refer to them in their planning and writing. Sharing models may be insufficient, however. Anne worries that students may not recognise what makes the work good and may fixate on punctuation or length: necessary qualities, but insufficient for success. "Simply providing students with contrasting cases does not automatically produce deep understanding and self-assessment," however (Lin-Siegler, Shaenfield and Elder, 2015, p. 533). Students need to consider what success looks like carefully:

> We cannot assume that simply being shown excellent work is enough to develop excellence. Something more is needed: either the complex task needs to be broken down, or particular aspects of a quality piece of work need to be explicitly highlighted and emphasised by a teacher.
>
> (Christodoulou, 2017, p. 42)

Version 5

Anne goes back to Berger's work, and notes how he and his students use examples:

> We sit and we admire. We critique and discuss what makes the work powerful: what makes a piece of creative writing compelling and exciting; what makes a scientific or historical research project significant and stirring; what makes a novel mathematical solution so breath-taking.
>
> (2003, p. 31)

Anne realises why 'student-friendly mark schemes' seemed helpful: students articulating the criteria for excellent work in their own words makes sense. Where she went wrong, however, was in copying the technique without understanding this: she gave students mark schemes, but students must begin with the models and, understanding them, articulate the standards they reach. Anne seeks tasks which will help students engage critically with the models and allow them to codify success in their own words. She identifies several powerful approaches:

5a) Comparing models

Students read examples of strong, average and weak paragraphs. This begins as a game of 'spot the difference', and then they highlight and describe strengths and weaknesses of each paragraph. Anne leads discussion of the strengths of a good paragraph; then she uses the weakest model as a completion problem, asking students to rewrite it using the strengths they have identified.

5b) Identifying criteria

Students read a strong and a weak experiment report. They compare them, section by section, and formulate what a strong report must include. This forms a checklist for them, but also leads them to return to the models as they write their own reports, rather than relying on the list of criteria alone.

5c) Comparing possible choices

Anne shows students problems, sentences or movements, and then offers a range of choices for the next word, number or movement. She asks students to choose what works next: which sentence will most elegantly conclude the argument, which word will complete the sentence grammatically, which line of working should come next. Anne gives students a hinge question (Problem 5) each time to check who has chosen what, and then asks students to explain the reasons for the choices they have made. Students get used to seeing each action and word as a choice and identifying the choices which work best.

5d) Examining improvements

Anne shows the process of improvement by sharing a weak paragraph and the same paragraph edited and improved. She invites students to identify the changes and explain the impact they have had. Then she asks students to apply the changes they have noticed to a new, weak paragraph, or to their own work.

5e) Live modelling

When Anne identifies an issue during a lesson and wants to remind students what success looks like, she sometimes models live. She revises and edits student work, or her own, on the board, asking students what her next move should be and why. Anne finds this a powerful way to give feedback to the whole class (Problem 6).

5f) Articulating success

After encountering the model, Anne asks students to record what they will remember about it. Sometimes, she asks them to codify this using a weak example: "What advice would you give to the person who wrote this?" Sometimes, she asks them to write goals for their own work. Anne returns to models across units; she asks students who are struggling or have made mistakes to return to strong models and identify what they have missed, or return to weak models and identify the traps into which they have fallen. Sometimes, Anne even has students draft answers first and then examine models; having begun the task, students can identify the choices made in the model better.

Experience - Michael Pershan

Helping students see what they can't yet see

Michael Pershan has been a math teacher in New York City since 2010, a few months after he graduated from university. It's the only job he's ever had, besides the little things over the summer when he was a teen: babysitter, camp counsellor, Pepsi vendor at a baseball stadium, tutor - all kind of relevant to teaching.

Though I teach math, math didn't feel easy for me as a student. It was never where I shined. An exception was geometry, with its heavy emphasis on proof. Proof felt natural for me in a way that algebra didn't.

When I began teaching, I realised that for many students the situation is reversed - it's proof that feels unnatural and cumbersome. Writing a proof involves combining statements in ways that seemed to mystify many students. This was especially true early in my career.

After a few years of hitting my head against the wall, I started to understand what made this such a difficult skill to teach. Proof is the closest that mathematics comes to writing, and writing itself is impossible without reading. How can a student who has never read an essay possibly write one? I concluded that my students needed to read more proofs.

It took me a few more years to understand how exactly to pull this off in class. My big frustration was that my students wouldn't devote enough attention to the proof examples I shared. I would distribute a completed proof and ask the class to read it with care. Very often, it seemed that they missed the whole point of the proof. They couldn't read it carefully yet – they didn't know how.

Now, things go better when I share proofs in class. One big difference is I have a much better understanding of all the subtle conceptual understandings that go into a proof, many of which were invisible to me at first. (In teaching, it can be trouble when a topic comes naturally to you.) There are many aspects of a proof that I need to help them uncover.

Besides a better understanding of the subtleties of the proof, I've learned to structure my activities in sturdier ways. I've learned to design these activities so that they have three parts:

1. The proof example
2. Comprehension questions about the example
3. Proof-writing practice, with the example as a model

I didn't come to this structure on my own, by the way. I came to it through reading about Cognitive Load Theory (where these are sometimes called "example-problem pairs") and especially from seeing it in some especially well-designed curricular materials. (In fact, I didn't really understand how to make my own example activities until I saw many models in these curricular materials. I needed examples, myself.)

So, for instance, I created the proof example seen in Figure 3.1 for my students this year.

Looking back, the example isn't perfect. It ended up being a bit visually crowded, and it might have been better to eliminate some of the letter abbreviations. In class, I actually covered up each stage of the proof to focus their attention on each part.

In any event, this activity shows a lot of what I've learned about teaching proof. I knew I wanted to make explicit the complicated two-stage structure of some congruence arguments, so I worked hard to create a pretty clear example for my students. I then called on students to answer a trio of analysis questions about the proof – there's a lot to notice, and students don't yet know how to notice the underlying structure of this kind of proof all on their own. Finally, I asked students to use what they'd noticed on a related pair of problems, so that students saw there was something that was generalisable to many different kinds of diagrams.

Even when my proof activities aren't structured so rigidly, I try to include variety and a chance to practice. Figure 3.2 shows a simpler activity, but I still call for students to do a bit of proof-completion in the second prompt:

Sometimes when I talk to other teachers about examples, they tell me they're worried that kids will just try to unthinkingly copy the model. I do know what they mean, but it's not what I see with my kids. I think that part of the reason is that I reserve example-analysis for when I worry that the math is going to be difficult, even overwhelming for many students. There is certainly a way to misuse these activities, and perhaps if I used these sorts of tasks on less complex material I would see unthinking imitation.

One of my jobs is to help students see things that they can't yet see – things like the logical structure of a good mathematical argument, or the way just a tiny bit of

Congruent Triangle Proofs | Name: ____________________

Example - *Tina proved this correctly. Go Tina!*

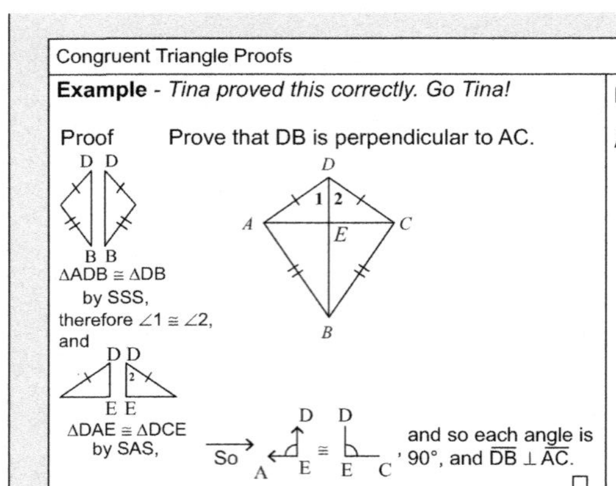

Analysis

- What two triangles did Tina first prove to be congruent? What info did she use to guarantee their congruence?
- What two triangles did Tina next prove to be congruent? What info did she use to guarantee their congruence?
- Why does proving that angle AED and AEC are congruent prove that the lines are perpendicular?

Practice

AB = AC, and AD bisects angle BAC.

a. Prove that D is the midpoint of BC
b. Show that AD is perpendicular to BC

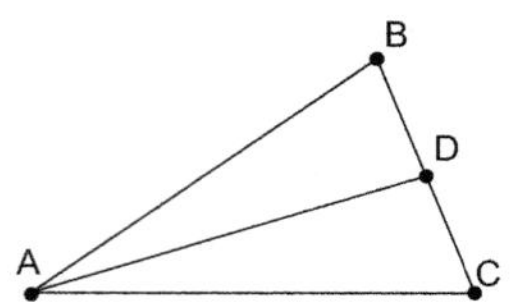

M is the midpoint of AC. Is it also the midpoint of BD?

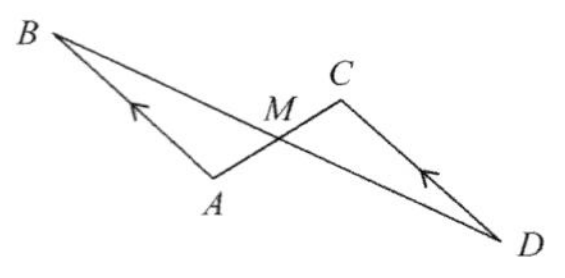

Figure 3.1 Proof example

Name: ____________________

$\overline{CS} \cong \overline{HR}$, $\angle 1 \cong \angle 2$

Is $\overline{CS} \cong HR$?

I, S, R, Y, C, 1, 2, H

Timothy's Argument

No, because if ∠ICR is wider there is a way to make $\overline{HS}$ longer than $\overline{CR}$ while keeping ∠1 ≅ ∠2 and CS ≅ HR.

Louise's Argument

Yes, because if you separate the overlapping triangles they are ≅ by SAS.

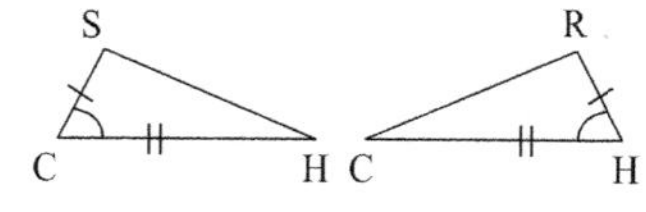

a. Where is the mistake in the mistaken argument?
b. For the mistaken argument to be true, what would the given information have to say?

Figure 3.2 Proof completion

information about a shape can guarantee a whole lot more. When things don't come naturally to my students, what I'm learning to do is to design an activity that opens up a little window into the mathematics so they can learn to see new things.

Conclusion

Anne has found ways to show students what success looks like. She tries to keep in mind that the criteria students articulate are simply descriptions of success; they can help students recall and articulate what they need to do, but they are abstractions from the models and cannot replace the models. She recalls the expertise reversal effect, too (Sweller et al., 2003): support which helps initially can hinder students once they have gained expertise. Early in the year, she helps students extensively in identifying the features of success; later, she expects students to identify these features with less support. Finally, she returns to the models in her feedback (Problem 6); now that students are clear what success looks like, Anne can highlight the gap between their work and the models, or have them identify that gap themselves. This deepens their sense of the features of success and their understanding of the quality of their own work. Having established for students what success looks like, Anne is keen to check what students have learned.

Checklist

1. Identify, obtain or create model work at different levels ☐
2. Identify key points students should be able to articulate, based on the models ☐
3. Design a task which will cause students to concentrate on the features of interest ☐
4. Prepare questions to further student thinking ☐
5. Consider asking students to improve weak models as completion problems ☐
6. Plan a way for students to record what they have learned ☐
7. Plan to return to the models later in the unit ☐

Double-check to avoid:

- Only showing excellent models ☐
- No opportunity to engage with the model ☐

A great read on this is . . .

Sadler, D. R. (1989). Formative assessment and the design of instructional systems. *Instruction Science*, 18, pp. 119–144.

Sadler explains how feedback and self-monitoring should work: the teacher sharing exemplars, leading students to experience and understand what leads to success and helping them to close the gap between the desired level of success and their own performances.

Note

1 This is also why I have offered both examples of successful applications of the principles and examples of near-misses throughout the book, in the hope of clarifying the principle.

4 How can we tell what students learned in the lesson?

The problem
It is hard to know what students have learned without a direct measure.

The evidence
We need objective measures of our impact
We should catch student errors early

The principle
Responsive teachers assess student learning at the end of each lesson and respond accordingly.

A practical tool
A task encapsulating the lesson, which shows what students have learned and allows us to respond.

Responding
Divide, dig and decide how to respond

Experience - Rowan Pearson, Jason Chahal
They are as much use for planning as assessment
Don't go it alone
Plan questions to identify gaps in student learning

Checklist

The problem

What students have learned is opaque without a direct measure.

Mike knows what he wants to teach, focuses lessons on that purpose and shows students what success looks like. But as students bustle out of the cramped classroom after a long hour and Mike slowly relaxes, he realises that for all his preparation and the students' hard work, he is unsure exactly what they learned. Students were busy: they practised the past tense in several ways. Yet he lacks anything answering the most important questions:

1 Did students really get it?
2 Are they ready to move on to the next lesson?

If he could answer these questions, he would be able to answer two other important questions, too:

1 Did my approach to this lesson work?
2 Am I improving as a teacher?

The evidence

We need objective measures of our impact
We should catch student errors early

The evidence suggests Mike needs to examine the impact of his teaching closely; he cannot rely on good intentions and common-sense approaches. He has judged lessons good if they felt successful, with smiling students and 'aha' moments, even if he overlooked a struggling student briefly; conversely, lessons seem failures if there is a challenging incident or a visibly baffled student, even if most students meet the objectives. Mike realises he judges lessons using 'poor proxies' – behaviour which is easy to observe, but does not mean students are learning (Coe, 2013, p. xii), such as:

- Students are busy; lots of work is done (especially written work).
- Students are engaged, interested, motivated.
- Students are getting attention: feedback, explanations.
- Classroom is ordered, calm, under control.
- Curriculum has been 'covered' (i.e. presented to students in some form).
- (At least some) students have supplied correct answers (whether or not they really understood them or could reproduce them independently).

Calm, work and engagement help students learn, but students could be calm, engaged and working and still learn nothing. Mike is surprised that even experts misjudge their impact; among 131 studies of feedback reviewed by Kluger and DeNisi (1996), fifty led to worse performance. He finds common-sense approaches which show negative results:

- **Scared Straight:** seeks to deter young people from crime by exposing them to life in prison; a review of randomised-controlled trials showed that those who participated were more likely to offend and reoffend than those who did not (Petrosino et al., 2013).
- **Infant simulators:** are used to deter teenage pregnancy by showing the burden of childcare. The first test of their effectiveness randomly-selected some girls to look after a 'virtual infant'. By the time they were 20, 17% of the girls who had cared for the virtual infant had become pregnant, as opposed to 11% of those who had not (Brinkman et al., 2016).

In both cases, objective measures made the harm caused by these plausible approaches visible. If researchers with time and knowledge can do more harm than good, Mike accepts that he cannot be certain about the impact of his teaching on students. A cynical friend puts it bluntly: teaching is really hard and most things do not work; it is better to assume everything is going to go wrong and find a way to identify what's gone wrong as quickly as possible than to assume everything is going to work.[1] Mike realises he needs to make what students are learning obvious by the end of the lesson, rather than assuming students' compliance and positive behaviour means they are learning.

Mike recognises that there are benefits and deterrents to examining the results of his teaching. He needs students to have mastered one lesson before he proceeds to the next, so they have the knowledge needed to understand the next topic (Introduction). Mike should know how successful his lessons were based on "how well they learned it, not how well [he] thought [he] taught it" (Lemov, 2015, p. 191); judgement based on his effort avoids examining the actual impact of the lesson (Coe, 2013, p. vii). Studies which have helped teachers better appreciate what their students are learning have inspired them to change their practice and increased student learning (Vescio, Ross and Adams, 2007; Supovitz, 2013). The deterrents are psychological: people see themselves in the best light (Ariely, 2013). Teachers tend to assume that their work benefits students, accepting student results they agree with (and which exceed their expectations) and discounting those which are worse than they expect (Lipsky, 1980, p. 71; Gipps, 1994, p. 31). Mike realises that despite the deterrents, examining the impact of his teaching could benefit his students.

- **Mike sees the benefits of objective measures of the impact of his teaching over subjective measures, and recognises the discomfort he must overcome to do so.**

Mike recognises that checking what students know at the end of a lesson risks conflating learning and performance:

- **Learning** is a permanent change in behaviour or knowledge.
- **Performance** is a temporary fluctuation in behaviour or knowledge which can be observed and measured during and immediately after acquisition (Soderstrom and Bjork, 2015).

Strategies to increase performance can hinder learning; strategies that decrease performance can help students apply knowledge better and retain it longer (Soderstrom and Bjork, 2015). For example, children were asked to make an equal number of throws:

- 2 and 4 feet from a target
- 3 feet from the target

Those throwing from 2 and 4 feet did worse initially; but in a subsequent test, *throwing from 3 feet*, they did better than those who had thrown from 3 feet originally: varying practice made the initial task harder, but increased learning (Kerr and Booth, 1978, in Soderstrom and Bjork, 2015). Similarly, completing sixteen maths problems in a random order is harder than when similar problems are grouped together; tested a week later, however, those who studied the randomly ordered problems answered three times more questions correctly (Rohrer and Taylor, 2007). David Didau claims that because performance and learning can sometimes be opposed, and because students may mimic desired answers, the effect of teaching on students' learning cannot be assessed meaningfully, nor can teaching be adapted accordingly (Didau, 2014; Didau and Rose, 2016, p. 102). Mike is aware that measuring what students know at the end of the lesson reveals temporary fluctuations in performance: "a highly imperfect index of long-term learning" (Soderstrom and Bjork, 2015, p. 188). Nonetheless, he wants to know what students know at the end of the lesson: he cannot rely on students remembering correct answers, but he can rely on incorrect answers reflecting gaps in their learning and can respond accordingly. He addresses students' forgetting by planning units to revisit key ideas (Problem 1). Conversely, genuine understanding may take longer than a lesson, but he cannot know that a lesson has been spent productively unless he checks; waiting longer than a lesson seems unwise, although he may check student understanding at milestones before the end of the lesson (Problem 5). Mike is conscious that learning and performance differ and that students will not retain what they know at the end of the lesson indefinitely, but he remains determined to identify student errors and misconceptions and respond rapidly.

- **Mike wants to check what students know at the end of a lesson to identify problems rapidly, not because they will remember the lesson's content for ever.**

The principle

Responsive teachers assess student learning at the end of each lesson and respond accordingly.

A practical tool

A task encapsulating the lesson, which shows what students have learned and allows us to respond.

To identify what students have learned, Mike needs an assessment which:

- Encapsulates the lesson's focus (based on Mike's solution to Problem 2).
- Students can complete swiftly (or would have completed anyway).
- He can examine swiftly (so he can use it every lesson).

Mike has not found research testing practical tools to do this: he sets out to design something which meets his needs.

Version 1: end-of-lesson tasks

Mike could use end-of-lesson tasks to assess what students have learned, such as:

- Write 3 things you've learned, 2 things you already knew, 1 thing you'd like to find out.
- What confused you about today's lesson?
- Where else could you use what you have learned today?

These questions are very open, however, and may not reveal whether students have understood the key points in the lesson. It may be worth knowing that "electromagnetic waves confused me" or "I learned I need to try harder," but this provides little guidance about where the problem lies, or how Mike should respond. Answers which reflect student confidence tell him even less about how much students have learned (see Problem 5). Mike realises that any end-of-lesson task will need to be more focused than these questions.

Version 2: examine student work

Mike may examine everything students have written during the lesson, or just the main task. This focuses on students' learning better than Version 1, but it is likely to take too long; two minutes on every student's book is already an hour's work. Even if he examines students' books, rather than marking them (Problem 6), Mike risks drowning in detail. It may not be obvious whether students have met the lesson's objectives; he may identify more issues than he can address. Rather than trawling through students' books, Mike would prefer to conduct "decision-driven data collection," designing tasks around what he wants to know (Wiliam, 2016, p. 108).

Version 3: use an aspect of the main task

Mike could assess just one aspect of students' work. For example, he might focus on:

- The introductory sentence to a paragraph.
- A single problem encapsulating the lesson's objective.
- A single, critical feature of a diagram or drawing, such as the correct labelling of axes.

This would allow him to check student understanding rapidly, using a task they were doing anyway, and respond accordingly. It would require no additional lesson time. However, he might struggle to find an aspect of the task encapsulating the lesson's purpose, and he would need to ensure students complete the task independently. When a question or task encapsulates the lesson, Mike uses it to assess what students have learned; where none exists, he designs an exit ticket.

Version 4: exit tickets

Mike can design an exit ticket to encapsulate the lesson's purpose in a task which can be completed and assessed rapidly. A good exit ticket should:

- Permit valid inferences about students' learning:
    - Differentiate accurately between levels of understanding
    - Elicit misconceptions
- Provide useful data:
    - Include everything that matters from the lesson
    - Focus on the key point
    - Have sufficient structure to elicit a clear response

- Be focused:
  - Swift to answer
  - Swift to mark

In English

VERSION 4A

Having taught students to use apostrophes, Mike's first thought is an exit ticket asking:

> When should an apostrophe be used?

It seems unlikely that students' answers will include everything they have learned, however. This invites a paragraph of disjointed ideas, slow to answer and from which Mike will have to glean what students have understood.

VERSION 4B

Mike considers:

Tick the sentences below if the apostrophe is correctly placed; put a cross if it is misplaced.

1. She's been hungry all day.
2. There are lots of dog's around here.
3. We were all very sad about Tim's illness.
4. Its Tuesday today.

This would be swift to answer and swift to mark, as it standardises the format of students' answers (Lemov, 2015). Mike may struggle to identify exactly what students understood, however; if a student ticks 'She's been hungry all day', do they recognise the contraction, or do they believe 'she' possesses 'hunger'? Moreover, the task does not demand much thinking; it would be hard to tell if students had guessed.

VERSION 4C

Mike seeks to retain a swift approach but to reveal student thinking better:

Add apostrophes to these sentences as necessary.

1. Hes very angry.
2. My dogs are unhappy.
3. Its on its way.
4. The ladies cars.

Each sentence reveals specific uses (and misuses) of the apostrophe students have understood.

In maths

VERSION 4A

Having taught students to add fractions with shared and different denominators, Mike considers asking:

1 $\frac{2}{7} + \frac{3}{7} =$

2 $\frac{1}{5} + \frac{2}{6} =$

This tells Mike whether students can add these fractions, but it does not show whether students can manage when the sum exceeds 1. Mike adds two more questions:

VERSION 4B

1 $\frac{2}{7} + \frac{3}{7} =$

2 $\frac{4}{5} + \frac{3}{5} =$

3 $\frac{1}{5} + \frac{2}{6} =$

4 $\frac{2}{3} + \frac{3}{4} =$

This allows him to distinguish whether students can add similar and different fractions, and whether they can manage both with totals that exceed 1. Mike could add more questions, but the benefits would be unlikely to outweigh the additional time needed.

Variations

Mike can design an exit ticket around factual knowledge, students' skills or their understanding. For example, in French, he could test students' knowledge of articles with an exit ticket asking students to:

Complete each sentence with the correct article.

1 J'aime bien __ foot.
2 Je voudrais __ stylo.
3 Je visite __ montagnes.
4 Je mange __ pomme.

He can test students' ability to apply their knowledge. For example, he may select three quotations from *Romeo and Juliet* relating to Romeo's character, such as Romeo's line: "But he that hath the steerage of my course, direct my sail / On, lusty gentlemen." He can test whether students can explain the significance of quotations in revealing Romeo's character by asking:

> 'Romeo is rash'. Explain how these quotations support this point.

Mike tailors exit tickets to the objective: he can use the same quotations to test different ways students may use their knowledge. If he wants to test students' skill in formulating points about Romeo's character, he may ask:

> Write a point sentence which uses each quotation to describe Romeo's character.

To test whether students can select evidence from the text to support points about Romeo's character, he can ask:

> 'Romeo is rash'. Identify three quotations supporting this point.

Finally, Mike can test students' understanding of concepts, for example:

> A solid metal object sinks: is this because of its weight?
>
> Why might someone confuse Hyperinflation (1923) with the Great Depression?

PREPARING EXIT TICKETS

Mike varies the amount of structure he provides students. Guidance seems to help students respond directly and concisely, so he:

- Formulates focused questions.
- Limits the space for student responses.
- Provides criteria, such as 'Include three points from the text in your response'.
- Uses multiple-choice questions as exit tickets (Problem 5).

Conversely, he values more open questions which require students to structure their answers: while students' initial answers to questions like 'Why did Henry VIII break from Rome?' may not be structured well, by providing feedback in the next lesson, Mike can help students understand how to structure answers and can then offer further practice. He balances the competing demands of structure and autonomy by gradually reducing structure as the year progresses.

Mike experiments with using slips of paper, printed sheets, tasks in students' books and online forms. Paper can be sorted more quickly, printed sheets allow greater structure, books leave a record and online forms can be marked automatically: he tailors his choices to the class and the learning.

When he can, Mike predicts likely student answers by writing an answer to each exit ticket as:

- A student with high prior attainment.
- A student with a basic grasp of the key ideas.
- A student holding serious misconceptions.

This establishes his minimum threshold for success, helps identify whether the exit ticket will elicit the desired response and shows how long students may need to complete it (Mike doubles the time it takes him). It helps Mike start planning to address errors and misconceptions and occasionally leads him to revise his lesson plan when he realises it fails to prepare students to complete the exit ticket successfully.

An opportunity for retrieval practice

Mike can use exit tickets to revisit previous learning. Testing past learning improves student retention significantly (Brown, Roediger and McDaniel, 2014; Pashler et al., 2007). Mike tries two-part exit tickets: a question from the current lesson and a revision question, based on his plan for revisiting past learning or to connect with a topic linked to the current lesson (Problem 1). Mike values the opportunity for retrieval practice but is wary of over-burdening the exit ticket or himself; often, he does not mark students' retrieval questions, but discusses them in class or has students assess them themselves.

Responding

Divide, dig and decide how to respond

Planning a response

Having collected students' answers, Mike wants to shape the next lesson using what they reveal. Five lessons using exit tickets could leave him with 150 slips of paper; two minutes on each would take five hours. Formulating the question carefully limits his task, but his response must be equally focused. Mike seeks to be utilitarian, seeking the greatest good for the greatest number, and mastery-oriented, helping every student reach a minimum threshold of understanding. To do this, he follows a three-step process: divide, dig, decide.

1) Divide

Mike sorts exit tickets into three piles:

- **Yes:** students definitely got it.
- **Maybe:** students partially understood/included elements of a good answer.
- **No:** students definitely did not get it.

Recording how each student did is onerous, but Mike finds that it proves valuable in checking students' success over time and examining how well students understand particular concepts.

2) Dig

Mike seeks to identify where students struggled. He splits the 'No' and 'Maybe' piles according to the errors and misconceptions they reveal, or studies five answers in depth to identify their characteristics. Examining the exit ticket for fractions, students have:

- Added the numerator and denominator separately $\left(\frac{5}{7} + \frac{3}{7} = \frac{8}{14}\right)$.
- Not found a common denominator, but used the larger number $\left(\frac{3}{5} + \frac{4}{6} = \frac{7}{6}\right)$.
- Not simplified the fraction $\left(\frac{3}{5} + \frac{4}{6} = \frac{38}{30}\right)$.

3) Decide

If every student answered well, Mike recycles the exit tickets and moves on. If every student answered poorly, he prepares to reteach from the start, recycling the exit tickets or returning them to students to correct. Usually, some students have got it right and some wrong. Mike can:

- Revise key points at the start of the next lesson:
    - Approaching directly, stating key points as fact: 'One common error was . . . it's important always to remember that . . .'
    - Approaching obliquely, reintroducing key ideas with fresh examples: 'Tell me what we can learn from this quotation about Romeo', using the representations collected in his unit plan (Problem 1).
- Model success for students (Problem 3):
    - Sharing a model student answer and discussing its strengths.
    - Sharing a partial answer and improving it collectively.
    - Sharing three answers and asking students to compare their strengths.
- Mark Yes/Maybe/No piles with a colour and ask students to revise or extend their answer, offering appropriate guidance to each group (Problem 6).
- Work with students according to their needs (Problem 6):
    - Sitting with students who struggled (and excelled) at a planned point in the lesson.
    - Setting students different tasks depending on the most suitable next steps.
    - Pairing students with different answers and asking them to compare the strengths and weaknesses of each.

For students' fractions exit tickets, Mike would:

- Model adding fractions with different denominators.
- Remind all students how to find a common denominator; ask a student who completed the exit ticket well to model this on the board.
- Highlight common errors while working through the example.
- Give five questions for practice:
    - One addition with shared denominators.
    - Two addition with different denominators.
    - Two subtraction with shared denominators.
- Check on students who struggled with the exit ticket during independent practice.

WHEN TO MOVE ON?

Mike must decide how much time to spend reviewing content. Should he accept that some students will not 'get it' and move on, or risk wasting the time of students who have 'got it'? The amount of time Mike spends reviewing the previous lesson depends on how important the key ideas are to the unit plan and when he plans to revisit them (Problem 1). There are no easy answers, but Mike uses the minimum threshold he has set for success to decide. Whenever possible, Mike wants to ensure every student understands the most important points, particularly in cumulative subjects like maths, in which students' progression depends on mastering each topic. Students who have 'got it' can be challenged to deepen their understanding, and they will have forgotten some of what they knew last lesson. Additional practice will do them no harm - indeed, it constitutes 'overlearning', or practice beyond the point of mastery, which significantly improves retention in the long-term (or, more accurately, slows forgetting; (Soderstrom and Bjork, 2015)).

RESPONDING IN PRACTICE

There are many ways to introduce the need to cover the previous lesson's work again. Mike may say:

- Not everyone put in a full effort yesterday.
- Since half of you didn't get it, we need to go over it again.
- You are going to do this properly, even if it takes all term.

Even if these statements are true, conveying exasperation or students' failings suggests frustration about a central feature of teaching. Students may be frustrated, too; if they are to engage wholeheartedly in revision and improvement, Mike needs to emphasise the value of the previous lesson's content, his faith they can succeed and his enthusiasm that they get the details right. Asking students to revisit maths problems, Mike tries:

> Your exit tickets were really helpful in understanding where we can all improve. I want us to go over an example of the harder kind of sums we were doing yesterday to pick up a couple of very important things to remember . . .
>
> [Later]
>
> Now, I have five questions to practice. All of you should find some of them difficult. Remember the three rules we've just discussed. Three minutes, go.

Experience - Rowan Pearson

They are as much use for planning as assessment

Don't go it alone

Rowan Pearson is Head of English at Phoenix Academy, a mixed school for 11-18-year-olds in West London. Now in his sixth year of teaching, Rowan is increasingly interested in teacher development - specifically in finding ways to make it easier for people to teach better.

What changed about your teaching?

Paradoxically, for a technique I initially thought would 'clinch' a lesson, or be the icing on the cake, I ended up seeing exit tickets as tools for planning as much as assessment. There is undoubtedly a blind spot when it comes to gauging pupils' progress, for all the reasons discussed elsewhere in this chapter. And undoubtedly exit tickets - when well-focused, linked to objectives and tightly limited in scope - offer a quick-fire measure of a lesson's outcomes and pupils' progress. But I found, in discussion with other teachers trialling the same technique, that the inevitable question once armed with a fistful of exit tickets was: what can I do with this data now I've got it?

The unanimous response to this question was to attempt to use exit tickets as a powerful bridge to meaningful starter activities in the next lesson. These starter activities wouldn't need to start from scratch - either in initiating new conversations, gauging prior knowledge or appealing for engagement - but would follow on seamlessly from the previous lesson. Using this technique forced me to think of lessons, even in the midst of mid-term survival mode, as sequences of learning, rather than standalone learning episodes. I had no choice but to use students' responses to directly address individual learning needs in my planning.

Now this may not exactly be revelatory. The benefits of such joined-up thinking are more or less a truism in teaching circles. In fact, medium-term planning and carefully detailed overviews of schemes of work exist specifically to achieve this kind of outcome. But nevertheless, my own experience of teaching has, to a large extent, been spent trying to address the disconnect between the casual idealism of pedagogical theory and the headrush-realism of everyday teaching practice. And for this reason, a technique which facilitates an easy transition from honest, efficient reflections on one lesson to a focused, impactful approach to a subsequent lesson has become a welcome addition to my own (modest) bank of teaching tricks.

What barriers did you face in using exit tickets? How did you overcome them?

I often find a major barrier to using new techniques with classes is getting some sort of buy-in from students. You need them to invest willingly in your shared work, and to be doing more than merely going through the motions. This can be tough. Over time I have found KS4 classes particularly, across the five schools I've worked in, to be increasingly outcomes-driven ('How is this helping me get an A?' etc.) and increasingly sceptical of change. Such a pragmatic obsession with results over learning can make students distrustful of things that go against their own tried-and-tested methods.

So first, I knew I had to make the point of exit tickets self-evident. I experimented with achieving this in a number of interconnected ways: by designing visually unique resources, by consistently utilising exit tickets and by celebrating students' work - and mistakes - in every lesson. I knew it wouldn't be enough to have students write their tickets in their books. It may sound trite, but if you are going to prize an exit ticket as a genuinely meaningful artefact of students' learning, and if you want any student to stretch themselves in the final throes of a lesson, having something visually different can help. (Also, it's quicker to flick through 30 sheets of A5 than 30 lumpen red books!)

But more than this, I made sure to show best, and sometimes worst, practice to the class in subsequent lessons. Suddenly, once students knew they may be quickly celebrated or held to account for their (lack of) work, the stakes were that little bit higher. This work will be looked at, and not a fortnight from now. In a sense, this smuggles some of the benefits of project-based learning into a more conventional learning journey. And thus simply handing out work from a lesson the day before becomes a very obvious marker for students that their learning is part of a continuing process.

What are your overall thoughts/tips on implementing exit tickets?

My main advice on implementing exit tickets would suit any new technique a teacher might trial – namely, don't go it alone! By this, I don't mean you need to joint-plan or team-teach your way through exit tickets, but having even one other person to discuss and reflect upon bumps and successes will make it far more likely that you will think meaningfully about how best to adapt exit tickets to your purposes. And it will also make it far more likely that you'll stick at it. This is because you will feel as if your own refinement of the process – whether the resources and activities themselves or the subsequent planning process – will be having an impact beyond your own classroom. And ultimately, this will always be an inspiring thought.

Finally, you may well move away from exit tickets for a while. But try before you do to write down your own 'golden rules' before putting techniques to one side – what worked for you? How did you overcome difficulties? What did students enjoy best? Even a few lines on these subjects will make it far easier to pick up where you left off in the future.

Experience – Jason Chahal

Plan questions to identify gaps in student learning

Jason Chahal is a primary school teacher who, over the past five years, has taught in coastal schools located in north Suffolk. Having taught across the primary phase, Jason currently teaches in KS2 where he has experienced the statutory end-of-KS2 testing regime. He has a keen interest in assessment and practical formative assessment techniques, which stems from reading Dylan Wiliam's Embedded Formative Assessment during initial teacher training. Determined to improve his teaching practice, he began to implement different techniques into his teaching and over time began to develop them in order to improve his use of formative assessment in the classroom.

I began using exit tickets around three years ago with an upper Key Stage 2 class that I taught for the last two years of their primary education. Initially, I began implementing exit tickets in maths lessons as I felt it was an easier fit in terms of writing questions & getting short, precise answers. What follows are the mistakes that I feel I made.

I saw them as a useful tool to use in observations to show how 'brilliant' my teaching was or how well the students had progressed within a lesson – a showpiece, if you will. This fitted with the 'rapid progress' rhetoric coming from Ofsted, school leaders and consultants alike. This was more the case at the school I was working at which, at the time, was expecting an Ofsted inspection. Of course, I wasn't a 'brilliant teacher' with pupils making 'rapid progress' in a lesson.

I quickly understood that their main use was to elicit the pupils' knowledge and understanding at that time in the lesson/learning phase. It dawned on me that the real value in these tickets was informing what to do in the next lesson. I often felt, to the dismay of my observers, that I found more use in identifying errors in pupils' responses rather than collecting a pile full of correct answers with streams of ticks on them.

I realise now that I was looking at learning as a process that happens over time, whereas I suspect that observers were still keen to look at lessons as a unit of time where learners made 'rapid progress'. However, I knew, having taught the class for two consecutive years, that the pupils could perform well in one lesson but six weeks later, on the same content, would have forgotten most of what was supposedly learned in the previous lessons. It seemed to me that the profession was hooked on believing that what is taught is learned and that the best teachers would be able to show this in a lesson. So, having a large pile of incorrectly answered exit tickets seemed like the efforts of an unsuccessful lesson and teacher when in fact what it showed was that the teacher recognised that students do not learn what we teach and so put into place an assessment strategy to identify what pupils' understanding was at that particular time in order to make adjustments for future lessons. As Dylan Wiliam (2013a) states:

> Our students do not learn what we teach. It is this simple and profound reality that means that assessment is perhaps the central process in effective instruction. If our students learned what we taught, we would never need to assess. We could simply catalog all the learning experiences we had organized for them, certain in the knowledge that this is what they had learned (p. 15).

As I was teaching a mixed-ability class, I began to differentiate the questions I would use at the end of a lesson when using exit tickets. The questions were all linked to the objective I wanted to assess but were tailored to what I thought the pupils would be able to answer. I had preconceived their understanding before I had even taught the lesson. Instead of posing a single question that could differentiate between different levels of understanding, I would write three questions that I'd anticipated each group of pupils could answer. This restricted the usefulness of the data obtained from the responses.

I believe that whilst my intentions were to assess pupils' understanding of what was taught in the lesson, by differentiating the questions three times, I was already eliminating what might have or have not been understood by the pupils in that lesson. Instead of differentiating by using multiple questions, I began to focus on a mastery approach by posing a single question and then asking the pupils to further explain to show a deeper understanding of what was covered.

I got lazy. I began to try and pose the exit ticket questions 'on-the-fly': I would try to generate the questions during the lessons by picking up on errors I could see walking around the classroom. This led to poor questions and responses that weren't allowing me to assess what I needed to in order to ascertain whether or not I could move on. It helped very little when planning the next lesson. My objective and questioning weren't tight enough and I knew it at the time.

An example from arithmetic (Figure 4.1): teaching the written method of subtraction by decomposition. The lesson focused on how to use the formal written method and follow the algorithm where they would be required to exchange/re-group where necessary. What I wanted to know by the end of lesson was whether the pupils understood the algorithm and could confidently use it in a variety of situations. I checked this by composing a question where there had been a successful exchange but made a common error where zero is used as a place holder and requires an additional exchange. I posed the following question:

Is this calculation correct? Can you explain your reasoning?

$$\begin{array}{rcccc} & 6 & 0 & \cancel{6} & 1 \\ - & 2 & 7 & 5 & 6 \\ \hline & 4 & 7 & 0 & 5 \\ \hline \end{array}$$

Figure 4.1 An arithmetic example

Potential responses with understanding:

- Pupil is unable to spot a mistake and therefore hasn't fully grasped how to use the algorithm in a situation where zero is a place holder. Possibly focuses on the initial exchange, believing that if that is correct then the whole calculation must be correct.
- Pupil can spot an error via carrying out the calculation themselves and spotting that an exchange is needed from the thousands column to the hundreds column. Writes the correct answer by using the algorithm to show their understanding.
- Pupil can spot the error and recognises zero as a place holder. Carries out the correct calculation and can articulate what error has been made and how to remedy the error using mathematical language such as place holder, negative number, exchange etc.

This allows me to distinguish between pupils who haven't quite mastered the algorithm and therefore can spot a correct exchange but can't see an error where zero is a place holder; can spot the error and rewrite the calculation correctly, thus understanding how to use the algorithm; and those who can spot the error, correct it and explain why someone might make this error.

I have now come to appreciate the value of a well-crafted question. As a result, I now know much attention, time and detail is needed to write a good question and even then, after using it in a lesson, there is always a period of reflection where I can see what tweaks are needed as a result of the pupils' responses. In the future, I aim to spend a good amount of time and thinking about crafting a good question.

I wouldn't always utilise the data gathered from the tickets and use them for the next lesson – I sometimes struggled with time management and workload or found it difficult to reflect in the next lesson with such widespread responses in a mixed ability class. I tried marking them (the exit tickets were stuck in their books) and then getting pupils to respond to the marking, which took its toll. As a result, I wasn't able to learn anything meaningful from them so that I could adapt my teaching for the next lesson.

There is little point in using exit tickets if I don't utilise the data derived from them. Responsive teaching would require me to use them in some way to shape the next lesson and improve upon what was taught and seemingly learned in the previous lesson.

Conclusion

We can identify what students have learned
We can improve our own work

Mike does not think exit tickets are the only way to identify what students have learned in a lesson, but he has yet to discover another approach which combines their efficiency and power. Exit tickets allow him to assess what students have learned rapidly and adapt the next lesson to fix gaps in their understanding sooner rather than later. Exit tickets also provide a mirror: they show the lesson as students experienced it, helping him to improve the clarity of his teaching and prepare for a more successful unit next year. After a few months' success with exit tickets, however, Mike becomes impatient. The end of a lesson seems too late; he wonders whether he can identify what students are thinking during the lesson.

A great read on this is . . .

Nuthall, G. (2007). *The hidden lives of learners*, 1st ed. Wellington, NZ: New Zealand Council for Educational Research.

Graham Nuthall went to amazing lengths to understand student learning. By recording every word spoken, written and heard by individual students during a term, and comparing them with teachers' goals, he demonstrated how easy it is to miss the complexity of students' thinking.

Checklist

1. Objective: do you have a clear lesson objective? ☐
2. Measure: is there an existing task which encapsulates success in the objective succinctly? ☐

   Yes: skip to 4.
   No: create an exit ticket, which:

   - Permits valid inferences about students' learning:
     - Differentiates accurately between levels of understanding ☐
     - Elicits misconceptions ☐
   - Provides useful data:
     - Includes everything that matters from the lesson ☐
     - Focuses on the key point ☐
     - Has sufficient structure to elicit a clear response ☐
   - Is focused:
     - Swift to answer ☐
     - Swift to mark ☐
3. Step back: are the objectives appropriate? Will the lesson prepare students to succeed? ☐
4. Assess students' work:
   - Divide: Yes/Maybe/No ☐
   - Dig: where have students struggled? ☐
   - Decide: how can their needs be met? ☐
5. Create the next step:
   - Utilitarian: seeking the greatest good for the greatest number. ☐
   - Mastery-oriented: seeking to get every student to a key level of understanding. ☐
6. Sell the next step: positive and mastery-oriented. ☐
7. Review your work: how can you teach this better next time? ☐

Double-check to avoid:

- Lengthy/vague questions you will not be able to respond to. ☐
- Asking questions without providing time to answer. ☐

Note

1 I'm indebted to Nick Hassey for this elegant formulation.

5 How can we tell what students are thinking?

The problem
It is hard to know what students are thinking, so they may maintain errors and misconceptions through the lesson.

The evidence
Monitoring student thinking about key points is essential
We need to know what everyone is thinking
We should check content, not confidence

The principle
Responsive teachers track student thinking to adapt teaching during lessons.

A practical tool
A hinge question which pinpoints misconceptions rapidly.

Responding
Plan first
Identify patterns
Lead discussion accordingly

Experience - Damian Benney
'Every wrong answer tells a story'
Identifying learning, not performance
Improving subject knowledge

Checklist

? The problem

It is hard to know what students are thinking, so they may maintain errors and misconceptions through the lesson.

Sara has clear objectives, useful models and an exit ticket ready, but she is unsure what students are thinking. She has explained the difference between hundreds, tens and ones, and asked individual students to contribute to her explanation. Before asking students to answer questions, she checked: 'Everybody happy?' Students answered with a couple of big grins, some vaguer smiles and a few averted eyes. Students began answering, but a few minutes later they seemed increasingly unsettled, and those she'd spoken to seemed confused. Sara stopped the class and checked what students understood. The more questions she asked, the clearer it became that many students had misunderstood key ideas. Sara had checked as she taught, but this had revealed little; the answers of those students happy to reply had obscured the confusion of others. Perhaps asking 'Everybody happy?' hinted that the correct answer was 'Yes': she was pretending to check, students were pretending to understand; she was not pushing students and they were not pushing her. Sara wondered whether this explained why misconceptions emerged in exit tickets which she had not spotted during the lesson:

- How could she know what students are thinking?

The evidence

Monitoring student thinking is essential
We need to know what everyone is thinking
We should check content, not confidence

> The central business of teaching is about creating changes in the minds of students - in what students know and believe and how they think. The ability to create change means that, in some way, teachers need to be constantly reading the minds of students. Are their minds focused? What are they understanding, or not understanding? What do they really think?
>
> (Nuthall, 2007, p. 23)

Sara can predict what students may learn, but she needs to monitor it closely; without knowing "what has changed in the minds, skills and attitudes of your students, you cannot really know how effective you have been" (Nuthall, 2007, p. 35). Exit tickets help, but only at the end of the lesson; Sara wants to track student thinking during the lesson and respond immediately. She wants to check student understanding while she explains, so she can clarify and modify her explanation; she wants to check whether students have understood key ideas before she asks them to apply them independently. Sara would love to check student thinking individually, but she cannot get around the class quickly enough. She needs the information she could get from an interview with each student in a fraction of the time it would take (Bart et al., 1994). Multiple choice questions and discussion of the answers seem to have a dramatic influence on student learning at university (Crouch et al., 2007). If Sara knew what students were thinking, she could identify the misconceptions she has predicted and respond with fresh explanations and representations (Problem 1).

- **Sara wants to respond to student thinking and misconceptions during the lesson.**

Track what everyone is thinking

Sara realises she may overlook some students. Benjamin Bloom (1984) noted that while most students benefited enormously from individual tutoring, one in five learned no more than they did being taught in a large group. He suggested that this derived, in part, from

> the unequal treatment of students within most classrooms. Observations of teacher interaction with students in the classroom reveal that teachers frequently direct their teaching and explanations to some students and ignore others. They give much positive reinforcement and encouragement to some students but not to others, and they encourage active participation in the classroom from some students and discourage it from others. The studies find that typically teachers give students in the top third of the class the greatest attention and students in the bottom third of the class receive the least attention and support.
>
> (Bloom, 1984, p. 11)

Students' gender, ethnicity and attainment significantly affect the likelihood they volunteer, or are invited, to contribute (Howe and Abedin, 2013). Sara judged whether students understood based on their answers, but the information she received was incomplete. Sometimes she nominated students whom she expected to answer correctly, knowing that a wrong answer would disrupt the lesson's flow; she avoided questioning students publicly if they lacked confidence; she spoke individually with students who seemed not to listen, or try, or get it. Unconsciously as much as consciously, Sara was judging student understanding based on how some students were doing; her teaching was responsive to their needs, not those of the whole class. Sara realised she could not identify how much the class understood based on the responses of the handful of students willing to volunteer answers (Wiliam, 2017). To respond to students' needs, Sara could not just search for right or wrong answers. She wanted to hear from every student and understand what they were thinking (Wiliam, 2011).

- **Sara wants to check what every student is thinking, not just those who are keen to participate.**

Check content, don't rely on confidence

Sara checked understanding by asking students to show how confident they were; their answers rarely helped. Sara wanted her students to be able to assess their own learning, but there are many conditions under which people misjudge how much they know (Koriat, 2007). In one study of patients who died in intensive care, doctors who were "'completely certain' of their diagnoses . . . were wrong 40% of the time" (Kahneman, 2011, p. 263). This suggests that

> it is wise to take admissions of uncertainty seriously, but declarations of high confidence mainly tell you that an individual has constructed a coherent story in his mind, not necessarily that the story is true.
>
> (Kahneman, 2011, p. 212)

Students are equally prone to overconfidence: "Without training, most learners cannot accurately judge what they do and don't know, and typically overestimate how well they have mastered material when they are finished studying" (Pashler et al., 2007, p. 23). Students lack "the necessary standards upon which to judge their learning state [and] the necessary knowledge to monitor their own state in comparison with the standards" (Kirschner and van Merriënboer, 2013, p. 177); they rely on inaccurate cues, like how they feel while learning, not how easy it is to recall something (Koriat, 2007). Those who know the least overestimate their knowledge most, because it is hardest for them to judge how well they are doing (Kruger and Dunning, 1999). Sara concludes that confidence is a poor indicator of competence.

Even if students could judge their learning accurately, their peers' scrutiny might deter honesty. Students may prefer to avoid looking stupid or delaying the class than to check with teachers. Few students would

> stop a group of twenty-five people and say, 'Uh, no. Actually, I don't really understand what you mean by the rigid structure of plant cells.' Even if they could, in that moment, (1) identify that there was something they didn't know, and (2) describe it quickly so that you could understand it, most people would be unlikely to do so in front of a group that size, out of embarrassment or fear that they would co-opt the better interests of the group. They'd assume they were the only one who didn't get it and that it wasn't fair to speak up.
>
> (Lemov, 2015, p. 30)

Sara may check students' confidence to adapt the lesson's pace and the support she offers, but she wants to avoid allowing questions of confidence, like 'Do students feel able to explain this process?', to take the place of harder questions, like 'Can students conduct this process?' (Kahneman, 2011, pp. 97–99). She can be guaranteed an honest self-assessment only if students know that they may have to justify their confidence; she will get useful information only if she hears students' explanations. So, Sara will check what students understand directly.

- **Knowing what students understand is more useful to Sara than knowing whether they are confident.**

The principle

Responsive teachers track student understanding during lessons, responding accordingly.

A practical tool

A hinge question which pinpoints misconceptions rapidly.

To track student understanding, Sara needs to:

- Discover what students are thinking rapidly.
- Hear from the whole class.
- Focus on content, not confidence.

Version 1: use windfall evidence

Sara responds to windfall evidence: spontaneous student comments revealing their understanding or misconceptions (Wiliam and Black, 1996). Such comments, to themselves or to peers, can be revealing, such as: 'You add first, then you do the multiplication', or 'Couldn't the king just do whatever he wanted?' Sara does not feel happy relying on windfall evidence, however, since students may be unwilling to voice their doubts, and she overhears only a fraction of what they say: she needs a more intentional approach.

Version 2: questioning for evidence

Sara uses questions to check student understanding. Teaching order of operations, she might ask:

> For $3 \times 4 + 2$, what will we do first? Why?

Answers will show whether students know multiplication comes before addition. They will reflect the class's thinking better if Sara selects respondents carefully, cold-calling, asking a range of students (Lemov, 2015, p. 36) or using lollipop sticks to choose at random. This only provides evidence for one student's thinking at a time, however. If it takes thirty seconds to get a clear answer from one student, it may take two minutes to hear from four; moreover, students' responses reflect each other's answers, masking their uncertainty. Questioning individual students is valuable in discussion and in modelling thinking, but it is too slow and inefficient for Sara to track the understanding of the whole class. Sara needs to hear from everyone.

Version 3: whole class checks of confidence

To assess the whole class, Sara might ask every student to give her a thumbs-up or down to show:

- How confident are you about what we've discussed so far?
- Are you ready to go on to the individual practice questions?
- Did you get the practice question right?

This solves one problem: it provides evidence for the whole class's thinking. It only tells Sara how well students believe they are doing, however - or how they want her and their peers to think they are doing - not how well they are actually doing. Sara wants to improve students' awareness of how well they are doing, but unverified confidence checks are unlikely to help. The only guarantee of an honest answer is ensuring students know there will be a check afterwards. Sara may ask two or three students to answer and justify their confidence, but this reintroduces the inefficiency of individual questions. Even honest answers may prove inaccurate: confidence is a poor indicator of competence. Sara needs to check students' understanding, not just their confidence.

Version 4: use hinge questions

Sara discovers an approach designed to show what students have understood: hinge questions. Hinge questions are multiple-choice questions where each answer option reflects an error or a line of reasoning, encouraging students to demonstrate their (mis)understanding (Parkes and Zimmaro, 2016). Ideally, students should be able to show Sara their answer within a minute and Sara should be able to assess them in fifteen seconds (Wiliam, 2011). By predicting and pre-empting possible wrong answers, hinge questions offer the rich information an individual interview would offer from the whole class (Bart et al., 1994). Many of Sara's colleagues are deeply sceptical about multiple-choice questions. This makes sense: they are used to seeing multiple-choice questions with answers which demand little thought. For example:

Which campaign was Martin Luther King involved in?

A The March on Washington
B The American Civil War
C Barack Obama's presidential campaign

Students with a basic grasp of chronology will recognise that only one of these events happened during Martin Luther King's lifetime. Such questions require little thinking and do not tell Sara anything useful about students' understanding (Smith, 2017). Sara does not want confirmation that students can avoid obviously wrong answers. She wants students to choose the right answer from plausible distractors: answers which seem logical but reflect misconceptions. A better question would ask:

Which campaign was Martin Luther King involved in?

A The March on Washington
B The Freedom Riders Campaign
C Lunch counter sit-ins

A is true, but B and C are plausible: Freedom Riders also campaigned for integrated travel on buses; lunch counter sit-ins were non-violent protests similar to those led by King. The incorrect answers are no longer silly mistakes; they reveal students' knowledge, reasoning and misconceptions. Answering the question helps students refine their understanding by reminding them of the variety of organisations campaigning for civil rights. Ordinarily, Sara hopes students will answer questions correctly; with a hinge question, she almost hopes students will answer incorrectly, as this will help her identify and respond to misconceptions rapidly. Sara can decide her next move based on a sense of every student's understanding.

Sara cannot find evidence that hinge questions have been tested rigorously in schools, but their use in universities has led to significant improvements in student learning according to a range of measures (Crouch et al., 2007). Sara thinks they could be applied productively to her classroom. She finds her experience mastering exit tickets (Problem 4) helpful in using hinge questions; while exit tickets leave her time to decide and prepare a response, hinge questions demand that she do so immediately. Designing the question carefully is essential.

1) Choose the hinge-point

Preparing a hinge question takes time, so Sara uses them sparingly, focusing on the most important idea in the lesson (identified in Problem 2). She places them at hinge-points, when students move from one activity to another: for example, near the end of an explanation, because students can only apply what she has explained if they understand it.

2) Design each answer to reflect just one misconception

Sara wants to know exactly what students are thinking. In particular, she wants to identify student misconceptions - beliefs which conflict with what is to be learned (Chi, 2008; see Problem 1). To identify exactly what students are thinking, each wrong answer choice should make sense to students who hold one - and only one - misconception. If students choose that answer, Sara knows which misconception they hold (and how to respond). She designs questions around the misconceptions collected for her unit plan: from external collections, experienced colleagues and from students' answers and comments (Problem 1; Gierl et al., 2017). Sara ensures each answer choice unambiguously reflects one student misconception. For example, testing student understanding of plate tectonics, the answer:

> Each continent sits on a plate.

does not show whether students believe there is one continent on each plate, or continents sit upon (but apart from) each plate. Better choices would be:

A Each continent has its own plate.
B Continents sit on top of (but separate from) plates.

Her final question would be:

Which of these is true of plates?

A Each continent has its own plate.
B Continents sit on top of (but separate from) plates.
C Oceans sit on top of plates (correct answer).

Similarly, testing order of operations, Sara's first draft question asks students whether the statement below is true or false.

> $22 - 5 \times 2^2$
>
> First I will subtract 22-5, working left to right.
> Then I will multiply 17×2^2.
> So I will do 17×4, which is 68.

This shows whether students understand the process, but if they are wrong, it does not show what they were thinking nor does it cover the range of misconceptions possible, such as using brackets. She refines the question to ask:

$(1 + 2 \times 3) + (4 - 2 + 1) =$

A 10 (the correct answer)
B 12 (students work left-to-right, rather than multiplying before adding, so they add 1 + 2 in the left brackets)
C 8 (students believe addition must always come before subtraction, so they subtract 4 - 3)

This helps identify two misconceptions, but Sara realises that if students combine the two misconceptions, they will reach the correct answer. She changes the numbers:

$(4 + 2 \times 3) + (7 - 2 + 1) =$

A 16 (correct)
B 24 (students work left to right, so add 4 + 2 in left brackets)
C 14 (students believe addition must always come before subtraction, so they subtract 7 - 3)
D 22 (students combine both misconceptions)

Sara ensures that each answer demonstrates a misconception unambiguously (for a discussion of how to do this which is as valuable as it is technical, see Wylie and Wiliam, 2007).

3) Balance brevity with covering likely misconceptions

Sara wants to include the most likely misconceptions while keeping questions succinct to allow rapid answers from students. The evidence suggests using only as many answers as are plausible. Three answers (one correct, two incorrect) seems best; adding additional plausible answers is fine, but adding implausible answers neither helps Sara nor challenges her students (Rodriguez, 2005). For the sake of brevity, Sara sometimes removes options: on plate tectonics, misconceptions include the belief that plates are separated by gaps, oceans or melted rock, but they all relate to the same fundamental idea, that plates do not touch one another. Sara might offer the answer choice 'plates never touch'; if students select it, she will then investigate what students believe separates plates. Other advice on writing effective multiple-choice questions includes:

- Put information in the question, not the answers (Haladyna, Downing and Rodriguez, 2002; Parkes and Zimmaro, 2016).
- Put answers which are dates and numbers in order; otherwise randomise the order of the answers (teachers often write the most plausible answer first) (Gierl et al., 2017).
- Keep answers similar in terms of content, structure and length (Gierl et al., 2017).

Sara reviews each answer for brevity, plausibility and the likelihood that students hold it before using the question.

Sara prefers to write and test her questions collaboratively with colleagues, in school or online. This ensures the wording is perfect, shares pedagogical content knowledge about misconceptions and creates a bank of questions (see Problem 7, Section 4 on making this possible). Writing hinge questions takes time, but Sara finds one question each lesson makes a significant difference. Sometimes, she uses existing problems and questions in the textbook, although it can be easier to start from the misconceptions than the questions. If she is short of time, she simply gets all students to respond to a question without having predicted their misconceptions; this weakens their usefulness but can still be revealing. She also uses students' comments in the lesson to create immediate checks; hearing a student's answer, Sara can stop the class immediately, and, without hinting at the answer, ask 'Michael said he thinks the answer is 7: hands up if you agree . . . Hands up if you disagree.' The more time Sara spends writing questions, individually and collaboratively, the better they are, but even a rough question to the whole class, formulated rapidly, offers better evidence of student learning than individual questions or confidence measures.

Testing conceptual knowledge

Hinge questions can test knowledge of concepts to a surprising depth. For example, Sara might ask:

Which of these was an immediate cause of Hitler becoming chancellor?

A Hyperinflation in Germany
B The death of Hindenburg
C The Reichstag Fire
D The collapse in Schleicher's authority

This tests chronological knowledge: two events (B and C) happened after Hitler became chancellor, one (A) happened a decade earlier. This is valuable, but Sara could test students' causal thinking more directly.

Which of these was the most immediate cause of Hitler becoming chancellor?

A Nazi violence intimidated many voters and opponents
B Schleicher's authority collapsed
C Hindenburg and Von Papen believed they could control Hitler as chancellor

Every answer contributed to Hitler becoming chancellor. Asking students to choose the 'most immediate' cause removes A from contention. A strong case could be made for B or C - this may launch further discussion. Asking students to prioritise causes, or consider their interaction, sharpens their disciplinary thinking.

Testing the application of knowledge

Including words like 'best' and 'most' in the question helps students think about what good answers look like. For example, Sara may wish to test students' ability to formulate a structured paragraph by asking them:

Which of these sentences best begins an answer to the question, "How does Shakespeare present Lady Macbeth's relationship with Macbeth in this scene?"

A Lady Macbeth sees Macbeth as cowardly and tells him what to do.
B Shakespeare shows that Lady Macbeth is in charge; for example, she tells Macbeth to "Give me the daggers."
C Lady Macbeth is shown as dominating Macbeth and acting decisively for them both.

All three sentences might appear somewhere within an answer; only C introduces the essay's argument. Question stems like 'Which of these sentences best begins . . .' can be used with any essay topic, helping students to grasp what a good introductory sentence looks like and limiting the burden of writing questions.

Testing definitions (and preventing guessing)

One concern about multiple-choice questions is that students may guess the answer. Sara can add further incorrect answers, as the example in Figure 5.1 demonstrates:

In which of these right-angled triangles is $a^2 + b^2 = c^2$?

(Wiliam, 2013b)

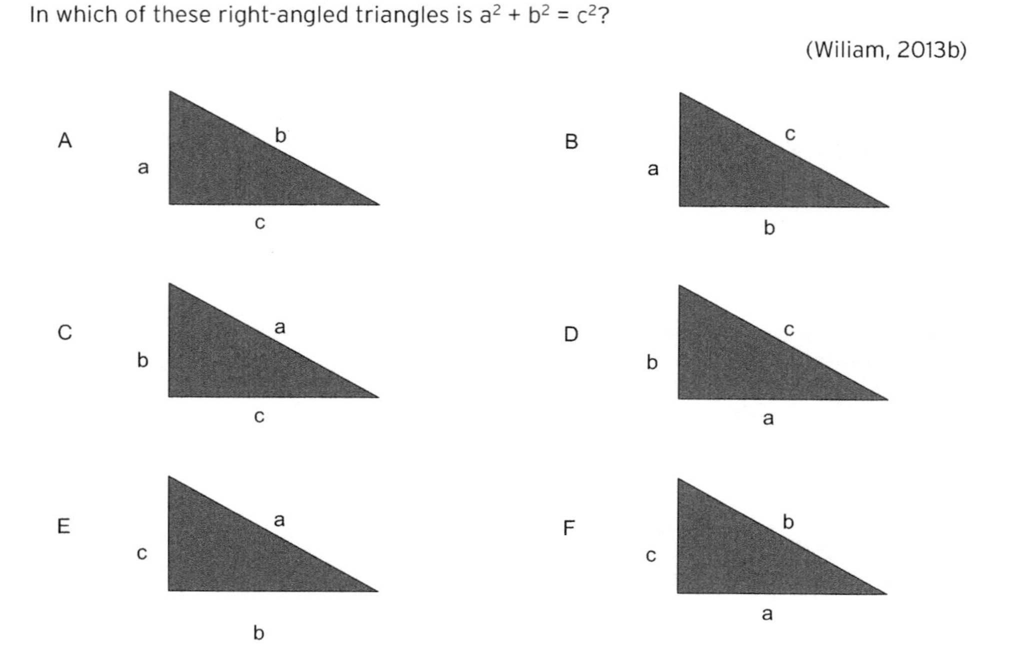

Figure 5.1 A hinge question in maths

There are two correct answers to this question (B and D); if students are unaware of this, they must consider all six triangles separately, which creates sixty-four possible answers, limiting the chances of guessing correctly (Wiliam, 2011). This should inculcate the habit of reading the question (and answers) carefully: if students stop reading at B, their answer is incorrect. While adding options is not recommended in general, as long as all the answers are plausible, as in this case, it seems justified (Rodriguez, 2005).

This example also demonstrates how a multiple-choice question can test conceptual understanding: one way to help students gain more flexible knowledge is asking them to make comparisons between different examples, identifying similarities and differences (Willingham, 2009). Students may have heard that $a^2 + b^2 = c^2$ in a right-angled triangle; a question like this demonstrates whether they have understood Pythagoras's Theorem. Comparing definitions, and choosing the most appropriate, sharpens students' understanding of a concept.

Sara finds she can write questions easily if she:

- Focuses on the lesson's key idea.
- Starts with misconceptions and formulates a question to elicit them.
- Re-uses questions across topics, changing the detail, keeping the structure.
- Collaborates with colleagues in writing, testing and refining questions.

Using hinge questions

Writing a hinge question is one challenge; Sara now needs to use them to assess students' understanding. Assessing students' thinking rapidly means asking all students to show their answers simultaneously, using whiteboards or by holding up their fingers to indicate their choice. Choreographing student responses and orchestrating the subsequent discussion requires effective classroom management. To get an answer from students within a minute and interpret that answer within fifteen seconds, Sara must avoid three pitfalls:

1) Invisible or illegible student responses

If Sara asks students to write on whiteboards, answers may come in a variety of scripts and sizes, with added decorations on waving whiteboards; if she asks students to raise their fingers (one finger is 'A', two fingers 'B'), students' hands may reach a variety of heights. Sara can:

- Ask students to minimise their response (just 'T' for 'True', 'A' for A).
- Clarify how students should write answers: 'Fill the whole board.'
- Say how the board or hands should be held: 'Two hands, shoulder height.'

This allows Sara to see every student's answer rapidly; it also means no student can be overlooked. Sara is tempted to ask students to show their working, but she recognises this provides an overwhelming amount of information and sticks to the answer, reassured that having predicted students' misconceptions, seeing students' working should be unnecessary.

2) Students copy each other Students will be tempted to change their answers if they see a peer's answers and think it correct. All this requires is a student to hold up their fingers two seconds early, or leave an answer visible on a whiteboard. Sara can ensure answers reflect students' own thinking by asking them to:

- Cover their whiteboards after writing their answer.
- Raise their whiteboards/fingers to a cue: 'All boards up in three, two, one.'
- Put their heads down or close their eyes before lifting fingers for an answer.
- Remember the purpose of the questions: helping students, not catching them out.

3) Calling out stilts discussion

Sara wants students to consider all answers carefully in the subsequent discussion. Students need to articulate their reasoning; those who were wrong need to understand why they were wrong, not just say they 'get it'. If a student derides others' answers as whiteboards are lifted – 'How is it A?' – other students are unlikely to explain, or even stand by, their answer. Alongside clear classroom management preventing calling out and mockery, Sara can convey the purpose of the task: reaching a common, improved understanding of the question and of errors to avoid. She may also teach students good discussion habits, such as ways to disagree politely and constructively.

4) Missing foundations and weak habits

Sara may benefit from making her expectations clear from the outset. Adapting the lesson while it is happening is one of the hardest things a teacher can do. Sara wants students to recognise that this is for their benefit and participate honestly, thoughtfully and patiently; this will be impossible if they believe she is trying to embarrass them. Sara explains her purpose when she introduces hinge questions and offers brief reminders of her aims and expectations frequently. When introducing hinge questions, and any time students need a full reminder, Sara may:

1. **Explain her goals:** 'I need to know what you're thinking, you need to think through the answers, so please give your own answer promptly, and listen respectfully to others.'
2. **Rehearse:** 'We're now going to practice with this question. First, write your answer, then keep it secret so that everyone can write their own, then we'll all hold them up on the count of three.'
3. **Do it again:** 'That was good, but I'd like to see everyone's board held up straight towards me on the three, so let's try that again, boards down . . . and on three: one, two three.' Rather than getting frustrated or accepting a poor response, 'do it again' solves the problem immediately and habituates students to success (Lemov, 2015).

Using technology

Sara can collate and record answers using technology. For example, she can design quizzes on Kahoot which students respond to on their phones. This allows her to choreograph student responses, maintain anonymity and make copying harder. It also provides Sara with a record of students' thinking, which she can compare with her record of exit tickets. Sara seeks technological approaches which make her approach faster and more efficient, and which allow her to keep students focused - she is naturally sceptical of encouraging student mobile phone use in the classroom. Nonetheless, she recognises that technology allows her to track student thinking more quickly and efficiently.

Responding

Plan first
Identify patterns
Challenge students to reconsider their ideas

The principles of Sara's response are the same whether she has used a hinge question, gained windfall evidence or identified students' thinking in some other way. As when she responds to exit tickets, Sara identifies prevalent misconceptions, gaps or errors, prioritises and responds accordingly. This may divert the lesson, but a misconception which will obstruct student understanding or is foundational to the subject is worth addressing immediately. Exit tickets leave Sara time to plan, but hinge questions demand an immediate response, so Sara needs to plan responses for the misconceptions she expects.

If students lack knowledge, Sara can respond relatively easily; changing their beliefs is harder, and the evidence on how to do so is mixed. Individual beliefs (thinking all blood vessels have valves) change with exposure to contradictory information, but mental models (such as a seeing the Earth as a flattened disc or hollow sphere) and category errors (seeing heat as an entity, rather than a process) are remarkably resistant to change (Chi, 2008). Presented with new, contradictory ideas, it is far easier to maintain existing mental models and fit new ideas into them (or ignore contradictory ideas) than it is to change them. Nuthall (2007) gives the example of a student who misinterprets an experiment examining a pencil in a jar of water. What she notices first is the pencil being magnified by the water, not refraction. Throughout the rest of the unit, she keeps focusing on magnification; she ignores information and observations which conflict with this. Classic theory on cognitive conflict suggests changing students' beliefs relies on their recognising a troubling anomaly and wishing to make sense of it; students need to see alternative ideas as intelligible, plausible and fruitful (Posner et al., 1982). For students to learn from an anomaly, rather than ignore it, the contradiction between their beliefs and the anomaly needs to be clear. It may help to contrast different mental models and create new categories which help students reconstruct their mental models (seeing heat as an emergent process, for example) (Chi, 2008). Showing the range of competing conceptions available may help: exposure to a range of questions in a variety of contexts - with plausible wrong answers - seems to be one way to encourage students to use correct conceptions. While misconceptions may always remain in people's minds and may "resurrect" themselves even after effective teaching (Potvin, Sauriol and Riopel, 2015, p. 8), Sara wants at least to challenge students to consider different conceptions and have the chance to change their mental models. Her approach is to plan, scan for patterns and lead discussions accordingly.

Plan first

Sara has predicted student misconceptions in writing the question, so she can plan responses for each misconception. To challenge students' conceptions, sometimes reiterating the right answer may be sufficient; usually it will not be. More likely, 'hindsight bias' will kick in for students - "When an unpredicted event occurs, we immediately adjust our view of the world to accommodate the surprise" (Kahneman, 2011, p. 202) - concluding that the answer was correct or the error inconsequential. Sara can:

- **Highlight anomalies:** 'If we believe X, how come Y . . .?'
- **Induce dissatisfaction:** 'X can't be true, because we can see that . . .'
- **Offer intelligible new conceptions:** 'Look at this different example.'
  This does not mean exhaustive preparation, but it is easier to identify a spare representation in advance than in the middle of the lesson. Sara's planning document (Problem 1) may help.
- **Introduce opportunities and challenges for debate:** 'Who would like to disagree . . .?'
- **Ask students to explain their reasoning and persuade one another:** 'What makes you think that?'

Scan for patterns, decide accordingly

As students respond, Sara scans their answers to decide what to do next. She takes a utilitarian approach, as with exit tickets (Problem 4), addressing the most common misconceptions and focusing on key ideas. If she identifies unexpected misconceptions (or they prove stubborn), she may wait to respond in the next lesson, having had time to formulate effective responses and using the representations collected in her unit plan (Problem 1). Occasionally, unexpectedly deep and complicated misconceptions emerge: the belief, for example, that most twentieth century world affairs were coordinated by the Illuminati - Sara may wait until the next lesson to marshal her arguments. If very few students get the answer correct, Sara reteaches immediately; if some students got the answer right, Sara moves to discussion.

Lead discussion

Sara helps students think through the reasoning by:

1. Highlighting specific answers to discuss, written on the board, shown on the visualiser or on a PowerPoint.
2. Asking students to explain how they reached their answers.
3. Eliciting and exploring misconceptions.
4. Helping students think through misconceptions to reach a logical, correct answer.

Usually, Sara maintains students' uncertainty by avoiding stating which answer is correct. She does not want students just to know the correct answer (and then stop thinking); she wants them to consider their reasoning carefully. Those who were wrong need to work out where they went wrong; those who were right need to be able to explain exactly why. When students are confident about their response – and they are right – they pay little attention to feedback (Kulhavy and Stock, 1989, in Hattie and Timperley, 2007). Derek Muller (2011) found that when a student is presented with correct expositions of scientific concepts:

> The student thinks they already know it, and they don't really pay utmost attention, they don't realise what's being presented differs from their prior knowledge, and they just get more confident in the things they were thinking beforehand.

Students think hardest while they are still uncertain: after choosing an answer and before knowing whether it is correct. Maintaining some confusion seems likely to help students remember the correct answer (D'Mello et al., 2014); it appears to represent another example of germane cognitive load (Problem 2). To achieve this, Sara:

1. Avoids disclosing the answer until students have reasoned their way to it.
2. Manages her 'tell' (Lemov, 2015): guarding against subtle hints and facial expressions, such as responding to a wrong answer with a frown and a 'Not quite'; instead, she responds to right and wrong answers alike with a smile, poker face and 'Other thoughts?'
3. Maintains her enthusiasm and interest: Sara is fascinated by what students are thinking and seeks to listen interpretively, to learn from students' answers, rather than evaluatively, merely seeking the right one (Wiliam, 2011).

For the question:

$(5 - 3 + 1) + (11 + 5 \times 2)$

A 24 (correct)
B - 20 (students believe that addition always precedes subtraction, so final operation is 5–25)
C 38 (work left to right throughout)
D 35 (forget that order of operations still works within the brackets)

Half of students chose A, seven chose B and six chose D. The discussion below exemplifies how Sara might respond.

Speaker	*Words*	*Rationale*
Sara	I'm seeing several people with B, Jamal, you said it was -20, what did you do first?	Emphasising that this is a common misconception. Not stating that it is incorrect. Starting with an incorrect answer to work towards correct may act as a kind of 'tell', but starting with a right answer and then asking students to undermine it seems to create deeper confusion than is productive.

(Continued)

(Continued)

Speaker	*Words*	*Rationale*
Jamal	First, I did the brackets.	
Sara	What did you do in the first set of brackets?	Asking Jamal to elaborate, rather than switching to another student
Jamal	I added 3 and 1, which is 4, then I subtracted 4 from 5, which makes 1.	
Sara	OK, and what did you do in the second set of brackets . . . Oh, Robbie's got his hand up, Robbie, what did you want to say?	Maintaining a level tone, not highlighting the mistake immediately hoping that another student will contribute.
Robbie	He did it wrong, miss.	
Teacher	Why do you think that?	Withholding the tell and the correct answer; forcing Robbie to explain reasoning.
Robbie	You should add before you subtract, so it's 5 minus 3 plus 1, making 3.	
Sara	Which answer do we agree with, 1 or 3, and why . . .?	Maintaining student thinking, withholding the correct answer, asking for explanations, checking whether students are following the discussion.
Anna	Robbie's answer is correct, because we should move left to right in the sum.	
Sara	When should we move left to right in the sum - Jamal?	Returning to Jamal to check whether he understands the misconception.
Jamal	When it's addition and subtraction, or multiplication or division.	
Sara	OK, so on the left-hand side, we've agreed that is should be 3, because 5 minus 3 plus 1 equals 3. What do we get on the right-hand side - Aaron, what did you get?	Choosing a student who chose D. Upbeat tone: showing interest in student responses.
Aaron	11 plus 5 is 16, and 16 times 2 is 32, then I add 3, which makes 35.	
Sara	Someone who chose A, why do you disagree with A? Rebecca, thank you.	
Rebecca	Multiplication comes before addition, so it should be 2 times 5, which is 10, add 11, which is 21, so the answer is 24.	
Sara	Who thinks Rebecca is correct? Thumb up for yes, down for no. Why do you think Rebecca's correct - yes, Aaron, you've changed your answer, what persuaded you?	Checking for understanding from the whole class. Showing students they are welcome to change their answers.
Aaron	Because BIDMAS means we have to do multiplication before addition.	
Sara	Yes, even inside the brackets, the same rule applies. No one chose C, can anyone spot the pitfall you avoided there?	Helping students identify other possible misconceptions.

Variations

Sara identifies other ways to discuss answers to hinge questions (or any other source of evidence of student thinking). These include:

- Asking students to identify similarities and differences between different answers (Stein et al., 2008; Willingham, 2009).
- Inviting students to look around the room at their peers' answers and change their answer, provided they explain why they are changing.
- Asking students to pair up with others who chose different answers and persuade one another of the merits of their answers; provided sufficient students (perhaps more than 40%) answer correctly initially, almost all students who change their minds adopt the correct answer (Crouch et al., 2007).
- Concluding by asking students to record key points in the discussion: 'What are you going to remember to do/not to do?'
- Asking the same hinge question again (which seems like it should be too easy, but often reveals continuing student misconceptions).
- Asking a second question testing the same misconception (the question above, for example, but with different numbers).

Rather than continuing the initial debate, for a powerful misconception Sara might offer fresh representations and examples to help students think through the limits to their answers. If students are assessing the motivations of characters in literature, Sara might offer them quotations from the text and ask them to reconcile the quotation with their answer. In history, she might share additional examples or return to the original documents in more detail: 'You've described Magna Carta as protecting ordinary men, let's go back to the document, what shows us that?' In science, Sara might show students another example: 'You've said an element is made of different atoms joined together: is bronze an element?' Each representation should help students appreciate how superficially plausible answers differ, and therefore which is correct. She uses the representation and misconceptions prepared in the unit planning document (Problem 1).

Extended discussions are crucial for students to develop accurate mental models, but Sara uses them judiciously. They can absorb a huge amount of time; she may move on, even if not every misconception has been addressed this lesson. She focuses discussions on the lesson's main objectives (Problem 2). If wrong answers are caused by careless errors or a lack of knowledge, Sara states the knowledge or the error. Finally, hard questions and confusion may prove good for learning, but they may also prove dispiriting: Sara seeks to ensure students are used to this kind of discussion, appreciate its purpose and recognise how it helps.

Other ways to use hinge questions

Writing good hinge questions is hard. But once written, Sara can use them both within the lesson - year after year, because student misconceptions are likely to remain similar when dealing with new concepts – and at other points in a lesson sequence. Teachers equipped with banks of questions have used them to:

- Test prior knowledge at the beginning of a unit or of a lesson.
- Promote discussion and debate among students.
- Check students' understanding of what a good model is (Problem 3).
- Test students' learning at the end of a lesson as an exit ticket (Problem 4).
- Test learning at the end of a topic.
- Provide feedback, ensuring students know what an improved answer looks like (Problem 6).
- Develop their own subject knowledge.

(Millar and Hames, 2003)

Developing a bank of multiple-choice questions with colleagues therefore has huge potential and could be combined with planning units (Problem 1). Sara refines the questions based on how students do; she edits the wording and removes answers which no students choose as they do not provide helpful information (Gierl et al., 2017).

Experience – Damian Benney

'Every wrong answer tells a story'

Identifying learning, not performance

Improving my subject knowledge

Damian Benney is a Deputy Headteacher at Penyrheol Comprehensive School in Swansea. He is responsible for developing teaching and learning across the school; it is his fifth year in this role, but his nineteenth as a science teacher. Being in the classroom is still his favourite part of the job, and he is committed to being a better teacher this year than last year. He saw the attraction of hinge questions immediately, realising that "every wrong answer tells a story."

When I first started using hinge questions, I was setting them in a back to front manner. I would plan a lesson and then try to tag a hinge question somewhere into the lesson. It was often very much a bolt-on rather than an integral and natural part of lessons. By tagging it into the lesson I was in danger of using them as a gimmick: just using them because they are good practice rather than using them to develop the learning in the classroom.

The main way that using hinge questions has changed my teaching is that they have made me focus on what the students should be learning (and how this may knit together with their prior knowledge) and what misconceptions and errors they may have. With this knowledge in mind I can plan my lesson, and my delivery, looking to minimise these misconceptions. My preference for hinge questions is for them to be multiple choice and to have at least one competitive alternative. They help to interrogate how my students are thinking.

Rather than plan the lesson and then design a hinge question, I now plan learning with hinge questions at the very heart; it makes me focus on what they are thinking and makes their understanding, and their internal schema, as visible as possible. My use of hinge questions has deepened my subject knowledge. It has made me focus on what and how students think. As my subject knowledge develops I am able to set better hinge questions. It has been very much a virtuous circle.

Using hinge questions successfully has also reduced my workload in terms of providing feedback to students. If I don't think of a hinge question to implement in a lesson, I may well find that I am giving the same feedback on a common error that I could have picked up and dealt with via a well-designed hinge question. This means my feedback can now be more focused on further improving a pupil's understanding of a concept, rather than fixing a common error that has fundamentally undermined their understanding of a concept.

A hinge question, just like every tool in the teacher's formative assessment armoury, has to overcome the thorny issue of performance versus learning. A common misconception in Key Stage 3 Science is that respiration is the same as breathing. When explaining what respiration is most Science teachers will explicitly state that it is not the same as breathing. If this explanation was then followed by this hinge question:

What is respiration?

A Breathing
B A process in living cells that releases energy
C Getting rid of waste products

then 100% of pupils (as long as they had been paying attention) would opt for a). This would be performance, not learning, and is just mimicry of the earlier explanation.

A good hinge question here would interrogate their thinking rather than invite mimicry. A better hinge question could be:

What is the link between breathing and respiration?

A There is no link. They are totally different processes.
B Many animals breathe in air and remove the oxygen needed for cellular respiration.
C Many animals breathe in oxygen. This is needed for cellular respiration.
D All living things breathe in oxygen and this is needed for cellular respiration.

The latter hinge question is far better as it doesn't reward pure mimicry. It also focuses on how today's learning fits in with prior knowledge. If pupils answer a) then I can pursue that. 'No link? Is there anything common between the two processes? Why do humans breathe?' If they answer c) then I can deal with the stubborn misconception that we inhale oxygen rather than air. If they answer d) then I can deal with the misconception that plants breathe.

Someone could have a strong argument that it would be far better to explicitly deal with the above misconceptions in the teacher exposition part of the lesson. This is a fair challenge, but you could argue that if you did then all pupils would answer b) . . . but potentially this just shows mimicry from the exposition. In some scenarios, it may be beneficial to allow hinge questions to reveal misconceptions.

On occasion it is best to have no options (though crucially all pupils must respond). After teaching kinetic theory, the best question to ask is 'what is in the gaps between the particles in a gas?' There are only two likely answers, air or nothing. This is a question where I would not give the options. I would try to find out what they think and infer how they think about kinetic theory.

I came to the realisation that well-designed hinge questions can get me as close to seeing what my students are 'thinking'. Learning may be invisible, but formative assessment, in particular hinge questions, can make their thinking as visible as possible.

Conclusion

Sara does not feel able to write a good hinge question every lesson. She ensures at least that her questions include the whole class, not just those who are confident to respond, and test what students know, not their confidence. She values being conscious of her bias towards keeping the lesson moving, not checking student understanding. When she cannot write a hinge question, she takes the fundamental idea, getting a check of the whole class's understanding at once, and puts a sentence on the board or restates a student's answer and asks every student to indicate their agreement or disagreement. With a strong idea of student understanding, she is keen to ensure that she can help improve her students' thinking; this leads her to think more carefully about the feedback she offers.

Checklist

1 Objective: do you have a clear lesson objective? ☐
2 Identify the hinge-point in the lesson. ☐
3 Design a hinge question, with answers which:
- Are plausible: not obviously wrong ☐
- Reflect just one misconception ☐
- Are succinct ☐

4 Prepare to explain what you want and create routines so that students respond:
- Legibly ☐
- Promptly ☐
- Respectfully towards one another ☐

5 Prepare follow-up questions/probes which:
- Keep students thinking ☐
- Elicit reasoning ☐
- Help students to reach correct answers ☐

6 Check student understanding again

Double-check to avoid:

- Too many, implausible or ambiguous answers ☐

A great read on this is . . .

Millar, R. (2016). Using assessment to drive the development of teaching-learning sequences. In Lavonen, J., Juuti, K., Lampiselkä, J., Uitto, A. and Hahl, K. (eds.), *Electronic Proceedings of the ESERA 2015 Conference. Science education research: Engaging learners for a sustainable future, Part 11* (co-ed. J. Dolin & P. Kind). Helsinki, Finland: University of Helsinki, pp. 1631–1642. www.esera.org/publications/esera-conference-proceedings/science-education-research-esera-2015/

Robin Millar describes a five-step process moving from a curriculum map to 'evidence of learning items' – diagnostic questions. He offers examples of how teachers have used these questions, the enthusiasm with which they've been met and the effect this has had on teaching.

6 How can we help every student improve?

? The problem
Providing effective feedback to students sustainably.

Evidence/practice
Don't start with feedback: start with the preceding problems
Tailor feedback to improve the task, student understanding of the subject or self-regulation, or to link those levels
Show students your high standards and your belief they can improve
Gradually increase student responsibility for creating and improving work
Target, standardise and limit marking
Students may be supported to improve without feedback

The principle
Responsive teachers help students improve their work.

Experience - Warren Valentine, Robin Conway

Checklist

? The problem

Providing effective feedback to students sustainably.

The problem

Marcus has reached the end of a five-period day. He faces one hundred and fifty books, in five neat piles. Despite his careful planning and checks for understanding during the lesson, each book will contain slips, mistakes and misconceptions. Each student will have interpreted the task differently, misspelled different words and reached a different point in a journey towards excellence. Marcus will see two of these classes tomorrow and wants to offer them guidance on how to improve. Several questions concern him:

- What feedback will move every student closer to their goals?
- How can he ensure students understand, act upon, and learn from feedback?
- How can he avoid students concluding that their work - or they - are rubbish?

- How can he avoid students becoming dependent on his feedback, and teach them to improve for themselves?
- How can he mark efficiently and effectively?
- What are the alternatives to marking?

Marcus turns to the evidence on guiding student improvement.

The evidence

Feedback is powerful but problematic. Deliberate practice demands "feedback and modification of efforts in response to that feedback" (Ericsson and Pool, 2016, p. 99). Feedback can "significantly improve learning processes and outcomes" (Shute, 2008, p. 154). However, while feedback is one of the most powerful influences on learning, its effects are highly variable. This is illustrated vividly by Kluger and DeNisi's (1996) review. They found the average effect of six hundred experiments was impressive and positive; astonishingly however, in 38% of these experiments feedback had a negative effect - it made performance worse. They concluded, somewhat equivocally, that feedback sometimes improves performance, sometimes debilitates it and sometimes has no effect (p. 254). Providing effective feedback is therefore a huge challenge: "if delivered correctly," it can improve students' learning and performance (Shute, 2008, p. 154); if delivered poorly, students may give up, reject the feedback or choose an easier goal (Wiliam, 2011, p. 119) at a huge cost to teachers' time. There is much research, but it suffers from "many conflicting findings and no consistent pattern of results" (Shute, 2008, pp. 153-154). Marcus begins exploring the evidence, hoping to refine his approach by applying ideas from the evidence to his subject, not believing he will find straight answers or iron rules.

Don't start with feedback

Marcus's first conclusion is that effective feedback relies on having solved the preceding problems in *Responsive Teaching*:

- Marcus needs "a concept of quality appropriate to the task" (Sadler, 1989, p. 121): he needs to know what he wants students to learn and what success looks like (Problem 1).
- Marcus must share what success looks like with students (Problem 3) so that they come "to hold a concept of quality roughly similar to that held by the teacher" (Sadler, 1989, p. 121); if goals are poorly defined or understood, students tend to be unclear how to use feedback (Sadler, 1989; Hattie and Timperley, 2007).
- Marcus must plan and teach effectively (Problem 2): feedback's power lies in addressing "faulty interpretations, not a total lack of understanding" (Hattie and Timperley, 2007, p. 82).
- Marcus can only offer feedback if he has designed tasks which provide insight into what students are thinking (Wiliam, 2017; Problems 4 and 5); designing tasks with feedback in mind allows a focused response.

Marcus realises that feedback adds little unless his teaching is effective; it is "not 'the answer'; it is but one powerful answer" (Hattie and Timperley, 2007, p. 104). There are many situations in which feedback may not be the most effective way to help students improve; in some cases, it may be unhelpful. If students are struggling or lack foundational knowledge, for example, feedback is likely to be less helpful than explicit teaching. Marcus needs to ensure his teaching is effective before providing feedback and to avoid seeing feedback as the sole solution to students' struggles.

- **Marcus resolves to review his teaching before providing feedback, and to reteach students explicitly if students are struggling or lack foundational knowledge.**

What feedback will move every student closer to their goals?

Marcus's choice of feedback will reflect the nature of gaps in student learning. An error may reflect one of several underlying causes:

> It could be a slip - that is, a careless procedural mistake; or a misconception, some persistent conceptual or procedural confusion (or naive view); or a lack of understanding in the form of a missing bit of conceptual or procedural knowledge, without any persistent misconception. Each of these causes implies a different instructional action, from minimal feedback (for the slip), to reteaching (for the lack of understanding), to the significant investment required to engineer a deeper cognitive shift (for the misconception).
>
> (Bennett, 2011, p. 17)

Marcus needs to identify what he wants students to change before he considers how to offer feedback. The limitations of focusing on how feedback is given are demonstrated by an experiment in which a school adopted comment-only marking (removing grades from students' work), in line with the evidence. Students who received comments seemed to make less progress than those who received grades, contrary to what the evidence would have suggested. The authors concluded that giving comments has no effect; it is what the comment says which matters (Smith and Gorard, 2005). First, Marcus has to identify what he hopes to achieve with his feedback; only then can he worry about how to convey this.

Marcus clarifies his goals by thinking about feedback as targeting different levels of change. These levels range from making changes to the task to changing the student's approach to learning. Several reviewers have created different frameworks: the framework

Specific	Concrete ↑	This task	How can I get this done? How can I make this better?
		The subject	How can I do better in tasks like this? What does it mean to be good in this subject?
	Reflective	Self-regulation	How can I manage myself to learn better? Who do I want to be?
General	Existential ↓	**Self-evaluation**	**How good am I?**

Figure 6.1 Different levels of feedback

seen in Figure 6.1 incorporates the work of Hattie and Timperley (2007); Kluger and DeNisi (1996) and Pryor and Crossouard (2010, p. 270).

This is a way to think about where feedback is targeted, not a suggestion that specific feedback is good and general bad (or vice versa). Marcus wants students to make changes at every level at some stage. Feedback's effect lies in helping students to focus on a particular level (Hattie and Timperley, 2007; Kluger and DeNisi, 1996). Marcus therefore seeks clarity about the levels, their merits and their disadvantages.

Improving this task

Marcus may want students to improve their responses to the current task. Teachers focus most frequently on helping students improve a specific piece of work by suggesting corrections or stating whether an answer is right (Hattie and Timperley, 2007). Marcus might suggest:

- Try that again, but this time hold your head up throughout the movement.
- Rewrite your answer to question 3 removing the brackets at step 2.
- Paragraph 3 needs more evidence.
- There is a problem with your answer to question 4.

(These examples, and those throughout, begin with what Marcus might say to lower-attaining students, who will benefit from more directive feedback, and move towards what he might say to higher-attaining students, who may benefit from and think more about less directive feedback; see Shute, 2008).

This can help students improve the current task, but its effects are limited; students are unlikely to be able to transfer what they learn about one task to another (Hattie and Timperley, 2007; Kluger and DeNisi, 1996; Shute, 2008). People struggle to recognise that they can use the solution to one problem to solve an analogous problem, unless they receive a hint to do so (Gick and Holyoak, 1980). Task feedback may also interfere with students' concentration if they are conducting elaborate tasks, learning complex tasks or seeking to follow rules (Kluger and DeNisi, 1996). It is more important that feedback improve the student than that it improve the task (Wiliam, 2017): Marcus wonders how he can offer more general feedback.

Deepening understanding of the subject

Marcus may want to help students deepen their understanding of learning and performance in the subject. Feedback on more general approaches to the subject may help students to identify and correct errors, use better strategies and process learning more deeply: this should lead to deeper understanding and better transfer to new tasks (Hattie and Timperley, 2007). Marcus might therefore give feedback applicable to a range of tasks, such as:

- Always underline key words in the question, then write a plan linked to them.
- Look back at the original question after each step to check you are on track.
- Reframe the problem as a diagram.
- Once you've completed a design, go back to the brief and see if you've met the goals.
- What do we always do first when we identify a problem with our work?

This should help students to understand underlying features of success in the subject; however, students may struggle to apply these features to the current task without specific prompts.

Marcus may also help students understand the subject itself, and the learner's role within the subject. He may help students to adopt disciplinary habits, or recognise what being a great scientist requires, by highlighting features of good mathematical thinking, scientific reasoning or historical questions:

- A good mathematician always checks their working.
- We've discussed how a historian develops their argument from many case studies: how would you structure an argument which draws on all the examples we've discussed today?
- Great artists steal.
- The questions you're asking are the kind which professors of English write whole books debating - so it's good to explore both sides of the answer before reaching a conclusion.

Again, students may need prompts to apply these general ideas to specific tasks.

Improving self-regulation

Marcus may want students to understand better how they learn. First, this means helping students self-monitor, recognising how well they are doing, what they know, and what is working; then it means helping them self-manage, planning and adapting in response to their self-monitoring (Hattie and Timperley, 2007). For example, one experiment increased students' accuracy in assessing their current knowledge and helped them gain higher grades; this proved particularly powerful for lower-attaining students (Casselman and Atwood, 2017). Marcus may help students to identify their current knowledge, skill and learning gaps - self-monitoring - and to think about how they can respond - self-managing:

- How did how well you did today differ from what you expected?
- What do you need to study more to improve in this area?
- Which strategies that you used today worked well? Why?
- What will you do differently during tomorrow's practice session?

Self-monitoring and self-management helps; if it verges into feedback about students directly however, it can have negative effects.

Self-evaluation

Feedback about students themselves is less effective than feedback focused on the task, subject or self-regulation (Hattie and Timperley, 2007). If students receive feedback about themselves combined with feedback about the task, they are likely to focus on themselves, which will distract them from improving their work (Kluger and DeNisi, 1996). Students like praise, and it may help boost self-efficacy, but most reviews find it has little or no positive effect on student learning (because it offers no useful information about how to improve). Even if praise increases motivation briefly, students may become dependent on it to keep learning; removing feedback later may then have a negative effect (Kluger and DeNisi,

1996). So, feedback aimed at students directly is likely to distract them from improvement, such as:

- You are a good/bad/indifferent student.
- You always come up with excellent answers.
- You've tried very hard at this.

Marcus focuses on helping students improve their work, rather than offering praise or personal criticism.

Moving between levels

More powerful than feedback focused on any one of these levels may be feedback which links different levels. Feedback at any one level will be insufficient for success and too much feedback at any one level may detract from performance (Hattie and Timperley, 2007). Students may struggle to apply specific feedback to new tasks; they may struggle to apply general feedback to specific tasks. Marcus may therefore offer feedback which links levels, helping students to recognise how feedback about one task may apply to others, or how they can self-monitor better, based on a deeper understanding of the subject. There seems to be a "powerful interactive effect" between feedback to improve specific tasks and feedback to improve strategies, processes or self-regulation (Hattie and Timperley, 2007, pp. 90–93). Marcus can therefore help students improve immediately and adopt useful strategies by linking feedback about specific tasks with feedback which develops a deeper understanding of the subject and self-regulation. The examples below show how feedback focused on one level can be combined with feedback at other levels:

One level	*Linking levels*	*Goal*
Correct Question 2, dividing before adding.	*Correct Question 2; remember to use BIDMAS.*	Linking task and process
Change 'its' to 'it is'.	*Remember to write formally in business letters: check for abbreviations.*	Linking task and process
Redraft this paragraph: include a quotation for each underlined statement.	*Redraft this paragraph: justify each claim you make using evidence from the text.*	Linking task and process
Remember the steps in creating an accurate graph.	*Remember the steps in creating an accurate graph: you have missed two.*	Linking process and task
Clearer explanation needed.	*Clearer explanation needed: describe the effect of this change.*	Linking process and task
Body position is important.	*Body position is important: hold your arms straight throughout the movement.*	Linking process and task
What are the limits to what the evidence allows us to say about this?	*Historians consider the limits to their evidence: how much can we say with certainty about this?*	Linking process and subject
Science often advances through testing anomalies.	*Science often advances through testing anomalies: why did this prove fruitful in this case?*	Linking subject and process
Great artists steal.	*Great artists steal: how did you use the examples we looked at to help you here?*	Linking subject and process

Marcus also finds this useful in ensuring his feedback provides concrete guidance to students; he replaces sentences like:

> Try to expand on your points with more thorough analysis of Macbeth's character.

with a concrete instruction which still conveys the underlying point he is trying to make:

> Offer a more thorough analysis of Macbeth's character by discussing his doubts as well as his determination.

Marcus may not *write* such lengthy feedback; doing so is slow and makes it harder for students to understand and respond. The important thing is that he remains conscious of opportunities to help students link feedback on one level with other levels, whether in writing, verbal comments or whole-class teaching.

- **Marcus resolves to be intentional about the level on which he is giving feedback and to make links between different levels where possible.**

How can he ensure students understand, act upon and learn from feedback?

Valerie Shute likens effective feedback to a good murder: it requires motive, "the student needs it"; opportunity, "the student receives it in time to use it"; and means, "the student is able and willing to use it" (2008, p. 175). Marcus needs to choose the right level for his feedback, but he also needs to ensure students can and do act upon it. He examines ways to ensure students understand, act upon and learn from feedback.

Understand

Students need to understand feedback to benefit from it. Marcus is not surprised to learn that vague feedback is unhelpful; he is surprised to learn that the more complicated and detailed the feedback, the less students benefit (Shute, 2008). Similarly, Marcus has always preferred guiding students with "facilitative feedback" to correcting them with "directive feedback," but directive feedback seems to be more effective, particularly when students are new to a topic, whereas higher-attaining students may be able to use less directive feedback (Shute, 2008). One experiment examined how likely history students were to accept and respond to feedback: students were more likely to understand the problems and respond when the feedback identified the location and possible solutions to problems; additional explanation made them less likely to respond (Nelson and Schunn, 2008). An extreme illustration comes from attempts to encourage people to learn more about their pension options; reducing a one-hundred page booklet to a single page led ten times more people to respond (Behavioural Insights Team, 2017). Marcus notes the need to be clear about what is to be done, particularly when students have less knowledge, and to limit his explanatory feedback.

In practice, Marcus focuses on offering clear, direct feedback using economy of language. Economy of language means making his purpose clear by using "the words that best focus students on what is most important, and no more" (Lemov, 2015, p. 414). Marcus can offer a clear goal by writing:

> Rewrite this paragraph including supporting quotations for each statement.

He can explain how and why this should be done verbally, allowing him to check whether students understand what he is asking them to do and why. Such explanations are also more efficient when offered to small groups, or the whole class, than to individuals. Marcus begins to revise his feedback, taking baggy sentences such as:

> I want you to have another go at that, and this time, when you're coming to your run-up, think about your weight.

and increasing their clarity and purpose:

> Try again; this time put all your weight on your left foot.

This saves time and helps students who struggle to process longer instructions. It also leads him to push students to return to models (Problem 3/below) and identify what is needed themselves. When students struggle to act on feedback, his first question for himself is: is the feedback clear enough?

- **Marcus seeks to make his feedback clear, concise and direct.**

Act upon

If students do not respond to feedback, it is hard to be sure they have benefited. Defined strictly, information provided to students is feedback "only when it is used" (Sadler, 1989, p. 121). Whatever level feedback targets, whatever form it comes in, students need to act upon it; a useful test is whether feedback is "more work for the recipient than the donor" (Wiliam, 2011, p. 129). This requires offering students a task to do, rather than presenting them with information alone (Wiliam, 2017). Even if Marcus thinks his feedback is perfectly clear, he has learned to check, rather than assuming students have understood. Marcus considers:

- **Checking for understanding:** sometimes, Marcus checks whether students have understood feedback by asking them to restate it, explaining what they can improve. He also uses hinge questions (Problem 5) to test whether students' understanding has changed through feedback, or simply whether they have understood the feedback and are ready to act upon it, rather than repeating errors.
- **Corrections:** students can correct errors; this works particularly well when Marcus asks them to focus on specific errors, strategies or parts of the work.
- **More practice:** an extension to corrections is asking students to complete similar problems with feedback in mind. This is obviously applicable in subjects like art, music and PE, in which students may repeat performances immediately, but it can be applied to any subject.

- **Redrafting:** taking 'more practice' further, students can redraft their work incorporating feedback. Ron Berger challenges students to treat their work as a craft; he challenges teachers to value quality of work over quantity:

  "What could you possibly achieve of quality in a single draft? Would you ever put on a play without rehearsals? Give a concert without practicing first? How much editing went into every book we read? Students in my classroom often take pride in their dedication to drafts: I did thirteen drafts of this cover, they brag" (2003, p. 90).

  Marcus can ask students to rewrite a paragraph or an essay, to redraw a diagram or to redo a solution. Improving work is more worthwhile than acknowledging feedback, and more satisfying: it forces students to understand and act upon feedback. It demonstrates students' capacity to improve and reach high standards powerfully: "work of excellence is transformational. Once a student sees that he or she is capable of excellence, that student is never quite the same" (Berger, 2003, p. 8). This experience of mastery (and reflection upon how it was achieved) develops student self-efficacy: confidence in what they can achieve (Bandura, 1982). Marcus believes it is important to allow students to perfect their approach to the task through redrafting, even for exam questions they will later complete under time pressure.
- **Marcus identifies an efficient and worthwhile improvement task for students to act upon.**

Learn from

Marcus wants to check whether students have benefited from feedback without creating excessive work for himself. He was unenthused by the fashion for 'triple impact marking' and 'dialogic feedback', in which he offered feedback, students responded, then he responded to student responses. No evidence exists that this kind of marking benefits students (Elliott et al., 2016). Marcus was unsurprised; he found this approach exhausting and struggled to believe that marking the same piece of work repeatedly achieved much. Nonetheless, he wants to check that feedback has worked, to identify how much further he can push students and to promote metacognition. This seems important: students can misinterpret feedback and follow-up tasks just as easily as the initial learning, and witnessing improvement should be satisfying for Marcus and his students. At its simplest, this means having students complete improvements during the lesson (Wiliam, 2017). Beyond this, Marcus seeks efficient, sustainable ways to ensure students are benefiting from feedback by:

- **Checking student work while students are making improvements:** Marcus circulates while students are responding to feedback, focusing on critical points and students who particularly struggled. He comments rapidly: 'Yes' to one student, 'Nicely put' to another, 'Reread the instructions' to a third. He limits his comments; he wants students to focus on the feedback he has written, so he avoids offering additional support too early. Sometimes, he groups students by their improvement task, so he can support them more efficiently and they can help one another.
- **Challenging students to identify how they have improved:** after making improvements, Marcus sometimes asks students to summarise what they have changed and

why. This develops their self-monitoring and self-management; it also provides a useful summary to which they can return.

- **Planning to revisit key points:** Marcus tries never to confuse learning and performance; he wants students to remember what the feedback encouraged, not just respond automatically and then forget about it. He plans to check what students have learned by revisiting key points and misconceptions in future lessons.
- **Marcus checks what students have learned from feedback when he can do so sustainably.** He does not feel guilty when he cannot do so sustainably. He is concerned by another question, however: how can he offer clear, directive, challenging feedback without triggering an emotional response from students who readily take feedback as evidence of their failings?

How can he avoid students concluding that their work - or they - are rubbish?

Students' responses depend on what the feedback is; they also depend on how students receive feedback (Shute, 2008). Feedback promotes learning when students receive it mindfully (Bangert-Drowns et al., 1991), shifting their attention and therefore their behaviour (Kluger and DeNisi, 1996). How students use feedback is affected by their willingness to seek and act upon it, their confidence in their work and whether they believe success relies on their actions or external factors (Hattie and Timperley, 2007). Marcus is unconvinced by 'Growth Mindset', particularly given recent, large-scale experiments which found no evidence for it (Bahník and Vranka, 2017), but he does not need to believe in 'Growth Mindset' to want his students to welcome feedback, believe they can improve and persevere when challenged. He has always sought to avoid practices which may undermine students' use of feedback: being controlling, interrupting or evaluating students (Shute, 2008). Additionally, he realises the importance of:

- **Avoiding giving students grades:** this distracts them from comments and diminishes their interest in the task; comments alone should lead all students to improve their work (Butler, 1988).
- **Never hinting students should stop trying:** Marcus avoids giving proxies for grades such as exam marking criteria, or telling students they have reached a particular level; he wants every student to feel there is a new challenge for them.
- **Avoiding social comparison:** he does not want students to focus on comparing themselves with their peers. He avoids praise and uses model work from previous years, rather than the current one.

Students' emotional responses affect how they react to feedback; Marcus wants to ensure students embrace feedback, rather than feeling threatened. He is intrigued by experiments in which teachers commented on students' essays as they normally would, but some students also received notes assuring them that "I'm giving you these comments because I have very high expectations and I know that you can reach them" (Yeager et al., 2014, p. 809). The results were dramatic for students with low trust in the school. In one study, African-American students who received the message were dramatically more likely to choose to submit redrafted essays than those who had not received it. In another study, resubmission was compulsory; African-American recipients of the message gained far better marks, having acted on the suggested edits. In a third, guidance about the meaning and value of feedback helped increase students' scores, leading to a higher pass rate on courses. The interventions also increased students' trust in the school and teachers. So Marcus presents feedback in a way which helps students accept it by:

- **Discussing emotional responses to feedback** and the value of separating these responses from useful information: 'When you get feedback, it can feel like you've done something wrong - I sometimes feel the same way too. It's always worth remembering the feedback is about the work - not you - and it's a way to do better next time.'
- **Emphasising why he is offering feedback:** noting high standards for students and a belief they can meet them: 'I'm giving you this feedback because I expect every student to solve every problem accurately, and I know you can do it.'
- **Emphasising why students are improving work:** 'It's worth redoing this, because when you incorporate these suggestions, your argument will be far stronger.'

Marcus separates helping students feel positive about feedback from using feedback to make students feel positive. Designing feedback to help students improve is distinct from using feedback to make encouraging personal comments such as: 'I'm so impressed with how hard you've tried and I'm delighted you seem to be really enjoying the unit at the moment.' This conveys neither useful feedback nor credible positive regard. Marcus builds self-efficacy through asking students to reflect upon their efficacy after successful experiences (Bandura, 1982). Confusing feedback about the task with feedback about the individual distracts students from improving without increasing their motivation or enjoyment (Hattie and Timperley, 2007; Kluger and DeNisi, 1996). Marcus therefore decides to:

- **Separate feedback from relationship-building:** in giving feedback, Marcus follows Laura McInerney's advice to "JUST STICK TO THE POINT . . . you can show the kids you care by your behaviour in the classroom" (2013).

Finally, Marcus seeks to create a culture in which students are accustomed to receiving feedback, recognise its value and respond to it willingly. He:

- **Clarifies expectations:** Marcus explains when students will receive feedback, the format it will take and its purpose.
- **Normalises feedback:** students receive some feedback, of some kind, frequently.
- **Models using feedback:** Marcus shows how feedback can help improve students' work and how he uses feedback from fellow teachers and students to improve his teaching.
- **Celebrates improvement:** Marcus shows previous students' redrafting to demonstrate how much improvement is possible; he encourages students to recognise their own improvement by explaining how they have used feedback to refine their work.
- **Marcus encourages students to welcome feedback and to recognise its value:** his concern is that student may become too accustomed to feedback.

How can he avoid students becoming dependent on his feedback, and teach them to improve for themselves?

Limiting feedback

The better Marcus's feedback becomes, the more positively students respond and the more he worries about their becoming dependent. The effects of feedback may rely on students receiving it on a continuous basis; if so, students' performance may diminish when Marcus reduces feedback or students move on to new classes (Kluger and DeNisi, 1996). Marcus decides not to offer feedback when students receive feedback from the task itself. Students benefit from thinking carefully about the task and learning its rules themselves; his feedback may detract from this (Kluger and DeNisi, 1996). This works in predictable environments (Kahneman and Klein, 2009); wrong answers become clear rapidly when speaking a foreign language or using computer-aided instruction; in a dissection or an essay, students are less likely to recognise the sources of their errors. Marcus avoids giving feedback where students can learn from the task.

Marcus learns that delaying feedback may limit students' dependence and increase learning. Sometimes, immediate feedback is useful; students who are new to a topic acquire knowledge faster with feedback and are better able to learn procedural skills like programming and maths (Hattie and Timperley, 2007; Shute, 2008). Less effective learners have fewer self-regulation strategies, depend more on external feedback and rarely seek feedback themselves (Hattie and Timperley, 2007, p. 94); giving immediate feedback avoids leaving them to struggle unproductively. However, delaying feedback may be effective with more complex tasks, when Marcus wants students to transfer learning from one task to another and with higher attaining students, who may be able to identify errors themselves (Shute, 2008). Leaving students to think longer themselves may both reduce their dependence and increase their thinking about the task. Finally, when students are practising a task to build fluency, immediate error correction may distract from learning and reduce students' automaticity (Hattie and Timperley, 2007); continued practice may allow them to solve problems themselves. Delaying feedback is so counter-intuitive that Marcus finds his colleagues sceptical that it is ever justified; nonetheless, he feels it is worth trying, even if only to test whether it can cause students to take more responsibility for their work. Marcus wonders what else he can do to increase student responsibility for improving their own work.

Sharing responsibility with students

Marcus is sceptical about self and peer assessment. His enthusiasm was sapped by repeated attempts which foundered on simple, stubborn problems. Students were often unsure of the qualities they were seeking in the work they assessed, and commented on the length or neatness of an answer, rather than its accuracy or elegance. Students often assessed their peers, not their peers' work, providing positive feedback because their peer was 'smart', for example; Marcus struggled to train students to give candid feedback kindly. Marcus is also concerned about the difficulties people have in assessing their own understanding accurately (Problem 5); students sometimes assess their work as 'excellent' and resist subsequent feedback. While Marcus could see possible benefits, he did not feel they justified the time and effort needed to achieve them.

Nonetheless, the evidence convinces Marcus to give self- and peer-assessment another chance. Sadler (1989) argues that the information students need in order to improve can be generated by students themselves, and that teaching often seeks to "facilitate the transition from feedback to self-monitoring." Ultimately,

> the indispensable conditions for improvement are that the student comes to hold a concept of quality roughly similar to that held by the teacher, is able to monitor continuously the quality of what is being produced during the act of production itself, and has a repertoire of alternative moves or strategies from which to draw at any given point. In other words, students have to be able to judge the quality of what they are producing and be able to regulate what they are doing during the doing of it.
>
> (Sadler, 1989, p. 121)

Writing of deliberate practice, Ericsson explains the transition:

> Early in the training process much of the feedback will come from the teacher or coach, who will monitor progress, point out problems, and offer ways to address those problems. With time and experience students must learn to monitor themselves, spot mistakes, and adjust accordingly.
>
> (Ericsson and Pool, 2016, p. 99)

If students recognise what success looks like and can monitor the quality of their work and improve it, this would reduce the pressure on Marcus and move students towards success and autonomy. Marcus wonders what he would need to do to meet these conditions.

Marcus realises he struggled with peer- and self-assessment because he introduced it too early. Marcus set out to 'do' peer-assessment for its own sake; students struggled because they lacked the 'concept of quality' needed to assess effectively. If students are to self-monitor, Marcus must solve the previous problems in *Responsive Teaching*: he needs to be clear what he wants students to learn (Problem 1) and set tasks which will show whether students have succeeded (Problem 4). Most importantly, he must ensure students know what success looks like and how to achieve it (Problem 3): self-monitoring requires "effective mental representations" (Ericsson and Pool, 2016, p. 99). Having met these conditions, students should be able to identify and close gaps between their current work and their goal. Students must be used to understanding and using feedback; if they are accustomed to welcoming, acting upon and learning from teacher feedback, they should be able to do the same using peer feedback. So, before Marcus introduces peer feedback, he assures himself he has solved these previous problems. Where he has doubts, he:

- **Reviews what excellent work looks like:** revisiting the models students have seen and asking students to explain their qualities and limitations.
- **Revises his expectations for using feedback:** reminding students of how and why they receive feedback and how to benefit from it.

Having established these conditions, Marcus wonders what practical approaches might work for self- and peer-assessment, and decides to begin with checklists.

- **Before Marcus asks students to assess themselves or their peers, he ensures students can identify success and act on feedback.**

INTRODUCING CHECKLISTS

Marcus is impressed to read how checklists are used to remind professionals of key actions. Having seen how pilots and surgeons use checklists to prevent avoidable errors (Gawande, 2010), he decides to use them to help students (Fletcher-Wood, 2016). Marcus has been frustrated by students' poor proofreading. Once students have been taught to leave a line between paragraphs, put a box around a diagram or check their working, they should be responsible for doing so. Marcus develops checklists to help students identify simple errors in their work, or their peers', and to improve their self-monitoring.

A simple presentation checklist includes:

- Title, underlined
- Date, underlined
- New paragraphs indented
- Capitals at the beginning of every sentence, full stops at the end
- Spelling and grammar accuracy

In maths, he includes:

- Date and title
- Question numbers
- Units
- Decimal points
- Calculations

Marcus expects students to complete checklists like these whenever they submit work. Occasionally, students do not take this seriously, giving their work a cursory glance or waiting for Marcus's more authoritative feedback. Marcus takes an increasingly firm line, returning work to students if they have missed obvious errors he is confident they know. Checklists reduce the time Marcus spends marking missing capital letters and simple calculation errors, allowing him to focus on more significant comments. Yet he recognises their limits in codifying what success looks like (Problem 3); checklists are powerful if they deal with superficial features or they remind students to complete actions they know how to do, but they cannot convey what quality looks like.

REVIEWING WORK AGAINST MODELS

Marcus revisits models so students can identify how their work compares to a model of success. Students need to judge their work against models and mental representations of success (Sadler, 1989; Ericsson and Pool, 2016); Marcus asks them to compare their work to model paragraphs, solutions or performances. Feedback as to whether an answer is correct or incorrect tends not to be useful: students need to see correct answers (Bangert-Drowns et al., 1991; Kluger and DeNisi, 1996). Marcus shows model solutions and asks students to identify whether their own work is missing any features included in the model; he plays compositions and asks students to note similarities and differences to their own. Sometimes, students diverge from models in intentional, creative ways; asking them to compare their work to models helps them recognise why this has been successful. At other times, comparing their work to a model allows students to identify missing steps or misconceptions themselves. Marcus wants students to recognise the similarities and differences of their work and models, so they can self-monitor effectively.

Often, Marcus structures students' comparison of their work and model work. Sometimes, he provides a checklist alongside a model; in each paragraph in an essay explaining the reasons for the Break with Rome, he asks students to include:

- A named reason, linked to the question
- Evidence for the reason
- Explanation why that reason affected Henry's mind
- A link back to the question
- Evaluation of how this compares to other reasons in the essay

He would not be confident asking his students to assess using this list alone, but using it with a model, they can check what a 'link back to the question' looks like. Structures like this help students to focus on important features of the work, avoiding overwhelming them by asking them to compare everything. Eventually, Marcus hopes students will internalise models of good practice, creating mental models which allow self-monitoring and remove the need for a checklist.

Once he is sure students know what success looks like, Marcus seeks to overcome other barriers to effective self- and peer-assessment. This means modelling, too: he models giving brief, constructive feedback frequently and reminds students explicitly what it looks like (and does not look like) before asking them to give and receive it themselves. He models acting upon feedback, showing how students can edit and improve work. He reminds students to return to models and previous feedback when they approach new problems; this helps them transfer what they have learned previously to new tasks. Marcus remains wary of investing too much time in self- and peer-assessment, however, or beginning it too early. He introduces them in a gradual, structured, limited way; he assures himself students know what success looks like and how to act upon feedback and monitors peer feedback and students' reactions closely. The evidence has convinced Marcus that students should be playing a greater role in providing and acting upon feedback; his scepticism remains, but it pushes him to provide the models and structures to ensure students are genuinely benefiting, through a better sense of quality and better self-monitoring.

- **Marcus ensures students know what success looks like, can review their work in a structured way and know how to respond.**

How can he mark efficiently and effectively?

Marcus thought that improving feedback meant marking; the evidence forces him to reconsider its importance. There is "a striking disparity between the enormous amount of effort invested in marking books, and the very small number of robust studies that have been completed to date" (Elliott et al., 2016, p. 4). Most studies are small-scale and/or focus on marking in higher education or English as a Foreign Language; most examine short-term impact, not long-term outcomes. Nonetheless, marking has become "disproportionately valued by schools and . . . unnecessarily burdensome for teachers," according to the Independent Teacher Workload Review Group (2016, p. 5), which recommended that marking should be driven by professional judgement and be "meaningful, manageable and motivating." Robin Conway (2017), at John Mason School, adds that

> this is not just an externally imposed problem. I have found myself adding more and more to my 'depth' marking over recent years; seeking to address literacy, give targets, identify what strengths the work shows, model effective answers and give directives for the application of targets . . . in short to make each piece of marking the perfect 'solution' to student progress. Too rarely have I stopped to think carefully about what the impact of each piece of feedback was, or which parts of this exhaustive process were actually the ones that best supported students' learning. When students made progress it felt irresponsible to tinker. When they struggled it felt dangerous to step back and reduce my input . . . so I generally added more.

While marking may have a powerful impact sometimes, Marcus begins by recalling some pitfalls suggested by the evidence on feedback in general:

- When students receive feedback from the task itself, he should not distract them with additional feedback.
- When students know what success looks like, Marcus should help them to take responsibility for fixing errors themselves.
- The longer and more complicated the feedback, the less likely students are to understand it and respond.
- Feedback is only feedback if students respond to the information Marcus provides.

Marcus realises that many common ways of marking do not help students improve. For example, noting that he has provided verbal feedback or writing verbal feedback in students' books adds no additional information helping students improve. Similarly, 'ticking and flicking' his way through students' books gives the illusion he has examined the work but tells students nothing. Finally, telling students that their work has reached a particular grade highlights superficial features of the mark scheme but does not develop their understanding of a great piece of work. The underlying problem is that Marcus is marking for many purposes. Sometimes he marks to help students improve; sometimes, he marks to show parents, managers or inspectors that he is doing a good job. This seems to underpin the obsession with recording on paper what needs to be in students' heads.

- **Marcus resolves to be clear about his purposes. When he wants to help students improve, he focuses on that and avoids wasting his time on cosmetic flourishes.** When he needs to demonstrate his hard work to others, he adopts approaches dedicated to doing so: he considers collecting examples of students improving their work from one draft to another, for example, which would demonstrate their progress and provide a useful resource for him in future years.
- **Marcus recalls the many circumstances under which individual feedback may not be the best approach and considers alternatives. He counts the opportunity cost that marking entails: an hour marking is an hour not planning better explanations, for example.**
- **When Marcus does mark students' work, he applies the evidence about effective feedback in the most efficient way.**

Over half of respondents to the government's Workload Challenge named excessive marking as particularly burdensome (Gibson, Oliver and Dennison, 2015). Marcus seeks to follow three principles to make his marking manageable, efficient and worthwhile:

Targeting marking

Students can produce more work in an hour than a teacher can be expected to mark (or a student can be expected to improve subsequently). Marcus examines a handful of student responses to identify common issues, then targets a particular aspect of the work, for example:

- The opening sentence of every paragraph
- The accuracy of the drawing of diagrams
- Three questions which capture every major misconception

Targeted marking is swift, but it has additional advantages. It provides a focus for students' improvement and for Marcus's reteaching and modelling next lesson. He applies the same principle to correcting spelling, punctuation and grammar, noting the first three errors, then only highlighting subject-specific vocabulary students will not see corrected elsewhere. Marcus does not tell students in advance how he will target marking, partly to avoid them focusing their efforts on that aspect of the task, partly because he decides the target having examined a handful of student responses. However, he does explain what he has chosen to do and why afterwards, to avoid students believing all unmarked sections of their work are perfect, or that their efforts on the whole piece are not worthwhile. Targeted marking saves time while supporting improvement.

Standardising feedback

Most student responses fall into a handful of categories: 'got it', 'partly' and 'didn't get it'. Rather than writing the same feedback dozens of times, Marcus standardises his feedback, for example, by:

- **Creating a sheet with common targets for frequent tasks** (like writing an essay in a particular genre), on which to indicate student successes and problems with ticks and crosses.

- **Developing marking codes and use them instead of writing words in fully:** in history, for example, Marcus uses 'Ev' to represent 'Evidence'; a tick and 'Ev' means the evidence is strong; a circle around 'Ev' means more or better evidence is needed. Marcus uses this abbreviation on every history essay he marks.
- **Grouping next steps for students into standard tasks:** Marcus finds he usually wants students either to:
  - Repeat the initial task with additional scaffolding or support
  - Develop or improve their answer
  - Go beyond the initial task with a new challenge

  Rather than writing the same thing fifteen times, he writes 'Target 1' on the work and displays the targets on the whiteboard to discuss with students and ask them to record/act upon as necessary.
- **At John Mason School in Abingdon, teachers develop 'Examiners' Reports', which collate their observations on a class's answers:** in subjects such as English and history, text summaries show strengths and weaknesses applicable to most students – students then have to work out which strengths and weaknesses apply to which of their work. In science and maths, spreadsheets show questions which students had completed well and poorly, alongside common observations (Conway, 2017).

When a student's answer is unique, Marcus provides unique feedback; for students who have performed exceptionally or have missed the point entirely, he offers personal feedback and support. Most student responses show similar strengths and weaknesses, however: Marcus cannot justify offering unique feedback for common problems when he can provide group feedback more quickly and effectively.

Reducing marking (increasing thinking)

The more complicated feedback is, the less students seem able to benefit (Kulhavey et al., 1985, in Shute, 2008). The more Marcus writes, the less students are likely to improve: verbal explanation of feedback is likely to be more useful, as it allows him to check student understanding as he explains. Marcus recalls the importance of economy of language and marks as concisely and simply as possible. He finds that sometimes he can make simple comments which challenge students to think much harder than his normal marking. For example:

- While telling students which answers are wrong allows them to correct those answers, telling students that 'One of these solutions misses the third step, can you identify which, and fix it?' forces students to examine their answers again carefully and spot the difference between work which meets the desired standard and that which does not.
- Rather than explaining every problem, Marcus can highlight where there is an issue without explaining what it is, then ask students to compare their work with the model and identify how to improve.
- Writing a comment for each student on paper, then providing groups of four students with their four comments, Marcus asks students to examine one another's work and identify which comment applies to which answer.

- For each error he finds, Michael Pershan writes an example that is related, but not identical, to the original problem on a sheet: all students receive the same sheet, and are asked to find an example related to a question they struggled with, and fix it accordingly (2017).

Marcus is careful not to frustrate students with these techniques. He uses them when students have the knowledge and the confidence needed to succeed; he ensures they are motivated and understand the purposes of the exercise. If they are not going to identify the improvements, he does not leave them searching vainly; his goal is to foster more thinking, while marking more efficiently.

What are the alternatives to marking?

Following the evidence on marking often moves Marcus away from individualised feedback towards feedback and improvement tasks targeted at groups of students. The logical conclusion is to find ways to offer feedback which do not require him to put pen to paper at all.

Feedback during the lesson

When he can, Marcus offers students feedback during the lesson. He follows the same feedback principles: he chooses a level of feedback and asks students to act upon it. He offers individual verbal feedback sometimes, but he struggles to spend sufficient time with each individual student to explain points thoroughly. After verbal feedback, Marcus checks students' understanding, asking them to explain it, or he asks them to begin revising their work and returns to check soon afterwards. Alternatively, he indicates the next step for students on their work, providing something students can refer to again once he has moved on. Group feedback is more efficient than individual feedback; he stops the class if he finds three students with a similar problem, on the assumption that there are more students in the same situation; even if this overestimates the number of students who are struggling, overlearning - continued practice and improvement, beyond the point at which students have mastered something - has powerful positive effects on learning and memory (Soderstrom and Bjork, 2015).

When planning feedback after the lesson, Marcus begins by examining student answers and errors, like Jo Facer (2016), who reads all her students' books once or twice a week, reading sixty books in thirty minutes. As he writes, he follows her example, noting problematic spellings, students' strengths, the main issues to improve and individuals who have done particularly well or poorly. Having read students' work, rather than writing on it, he uses other strategies:

Reteaching

Reteaching allows Marcus to challenge common misconceptions or knowledge gaps efficiently. He reiterates definitions or offers mnemonics to support students' factual knowledge; he offers examples, counter-examples and big pictures to support conceptual understanding (Shute, 2008). He repeats initial teaching, using fresh images, examples and metaphors from his unit plan (Problem 1). Students who struggled to add using a number line may do better with counters; those confused by their reading about the American Constitution may benefit from studying current cases in the Supreme Court. Students who 'got it' last lesson need not get bored; they often forget aspects of the lesson, they can be asked to overlearn, to explain points to the class or they can be challenged with fresh tasks. Reteaching seems the simplest and most efficient way to approach knowledge gaps and misconceptions without giving individual feedback.

Revisit model work

Closing the gap between students' performances and goals may require more (or clearer) knowledge; it may also require clearer goals. Just as Marcus revisits what he has taught students, he revisits the models he offered, or provides fresh ones; students can now compare their work with the model and better understand where the gap lies. Carolyn Massey (2016) used Orlando Figes's *A People's Tragedy* to model good historical writing for her A-level students; she then sent students back to this model, asking them to examine pages which demonstrated what they needed to do: 'Your sentences are overlong, reread page 46.' Revisiting checklists (discussed in Problem 3) helps students identify missing features of their work: punctuation, point sentences or balanced equations. Usually, Marcus chooses the models students will examine rather than asking them to use one another's, to ensure they can see the features which seem most important. Revisiting goals allows students to improve their work and understand better what success looks like.

Revise the process

Marcus reminds students how to create a good piece of work. He models the process of improvement, providing demonstrations and worked examples to show what students can do to their work (Shute, 2008; Problem 3). For example, he takes a student's answer, or a weak example he has created himself, and models rewriting or correcting it on the board. He takes a paragraph and works through it line by line, reading the line and then asking students open questions initially, moving to direct questions if they struggle to spot the points he is making:

- How could this be improved?
- How could we put this more clearly?
- What specific term should we use here?
- Who can suggest an unnecessary word we could remove?

He asks students to compare weak examples with model work and identify what's missing: 'Go back to the model: which step is missing from the solution?' Demonstrating how to improve work and guide student thinking about the choices made has two effects. First, it reinforces students' understanding of the choices and decisions which help create great work, particularly when Marcus refers to the original models students saw (Problem 3). Second, it models the process of correction, editing and redrafting he wants students to use to refine their own work.

More practice

Knowing how much students know is important, but it does not mean Marcus has to intervene immediately. Students may benefit from further practice, perhaps even without error correction. Building on the distinction between learning and performance – that lower performance can sometimes lead to greater learning (see Problem 2), Josh Goodrich (2017) notes that teachers skilled in formative assessment can use this to keep tight control of student learning, mistakes and misconceptions. This can mean that students never get the chance to struggle, as teachers address misconceptions immediately without allowing students to do the thinking which may lead to longer-term learning. This is supported by Kluger and DeNisi's (1996, p. 265) observation that feedback "may reduce the cognitive effort involved in task performance" and so be "detrimental in the long run": there is value in allowing students with high prior knowledge to identify improvements themselves. As Goodrich observes, if teachers do not allow students to struggle, it can appear that students are doing well but may harm their longer-term retention. Deciding exactly how much extra practice to ask students to do is a matter of judgement, but Marcus tries leaving students to continue for at least two or three more problems or sentences, then checking back with them. Marcus finds this is not an easy message to convey – particularly to observers – but it is an important one: rapid feedback, particularly after students have acquired the knowledge they require, may diminish learning. Sometimes more practice helps most.

Conclusion

While it is helpful for Marcus to consider each strategy for guiding improvement individually, in practice his responses are far more fluid than his categorisation suggests. Over a handful of lessons, students may receive verbal feedback, targeted marking and feedback addressed to the whole class. Individual feedback may be supplemented by revising model work as a group, for example. Marcus shifts between strategies depending on the needs of individual students, just like Susan Strachan (2017); normally, she offers whole-class feedback, but if individuals will struggle to identify their own targets:

> I will pop their name next to a specific target and when I am circulating the class after giving the information to the students I will tell them what I want them to work on, or in the case of some of my weakest students I will have written their target quickly into their book.

Combining approaches allows Marcus to be efficient, while ensuring every student gets the help they need; it allows Marcus to give collective feedback, while ensuring students know how they can improve. Marcus finds avoiding individual feedback saves time he can use better. Instead of trying to explain how to improve work for every student in writing, he can spend time planning a five minute explanation which will show every student what he means clearly. Marcus's review of the evidence and his options convinces him that Shute is right to argue that there is no "best" type of feedback (2008, p. 182), but when he can follow the evidence and do so efficiently, he finds feedback can support students to improve powerfully.

The principle

Responsive teachers provide students with clear feedback tailored to what they need to improve, in a sustainable way.

Experience – Robin Conway

Saving time through group feedback

John Mason School is a comprehensive academy in Abingdon, Oxfordshire. Robin Conway has worked there as a teacher since 2007. A history teacher originally, he also teaches sociology, politics and psychology to A-level. He has worked as Head of History and Humanities and Professional Tutor, and he became Director of Research and Innovation in 2015 with a brief of supporting the school to bridge the gap between educational research and classroom practice. He writes for and edits the school's reflective blog (https://jmsreflect.blog/), from which this is an extract.

"At JMS we did indeed pilot this model of feedback across various subjects and key stages in order to reflect on the purpose of feedback and the impact it could have. There were a lot of positives to it: once teachers got into the swing it was a dramatic workload-saver. It drew my attention to exactly how much time I spend rewriting the same comments on several students' work. Instead, using this model we produced a single class feedback sheet, which we started terming the 'Examiner's Report' and then focused on how we would ensure that students took the key messages on board. As with any feedback model, simply telling students what had gone well and what needed improving was not enough. Modelling helped but even combined both methods rely on students being able to identify which aspects of the general feedback applied to their work. Those with lower confidence had a tendency to be over-critical of their work and risk focusing on fixing problems which did not apply. Those with a limited grasp of the assessment criteria could not always see which bits of feedback applied to them.

"One-to-one conversations with those students who struggled to apply the feedback were crucial. I think our openness that we were trying something new and wanted their feedback on it also helped; students seemed more willing to admit early on if they were struggling to understand the feedback. This may be because 'problems' could be safely located with the 'new' model, rather than in themselves or the teacher, which facilitated questions and dialogue.

"For me, the process has given a new emphasis to the importance of dialogue in feedback. I am not advocating extended written discussion, or even a specific pen colour. Workload has to be a consideration, but so does turnaround time if the effort is to pay off for the students. However I am convinced of the value to my students in seeing the feedback I give as the first step in a dialogic process where we discuss what went well and how that was achieved, what the next steps are and how they will try to meet these and then a way forward.

"This does not have to be a laborious written dialogue built in different colours over several weeks, with books and folders passed back and forth. Sometimes, often, verbal discussion is quicker and more directly relevant to the student or small group with whom I wish to discuss their work. Tools such as the 'examiner's report' marking can play a valuable part in this by cutting down wasted time marking repetitively whilst shaping my thoughts on how to move students forward and giving us a clear starting point for dialogue beyond the piece of work itself. However I have found whole-class feedback to be very much the start of a process, and not sufficient on its own. In whatever form I need my students to respond directly to my feedback to be sure that it is doing the job.

"Questions that helped me to reflect on student responses to feedback

1. How widespread is this error and is it something I need to address with the whole class?
2. Is this something the students can fix themselves? If so, when am I going to give them time to do that?
3. How will I know if this feedback has 'sunk in'? What am I expecting students to do with it or how am I expecting their thinking to develop? When am I going to give them time to do that?
4. What is the most time efficient way to work with the student on this development point?"

Experience – Warren Valentine

Planning the work which requires feedback

Warren Valentine is head of government and politics at a state grammar school in the southeast. He has been teaching for five years and has recently completed an MA in History Education. He has recently launched a working group in his school on feedback and assessment and continues to experiment with his practice in this area.

"I radically transformed my teaching practice when I experimented with looking at every student's book at the end of every lesson. I had grown frustrated with the way I had bought into a system that conflated the terms 'marking' and 'feedback' and mandated that some form of 'formative feedback' be provided to students once every six lessons. This ultimately led to students working towards an 'outcome' piece of work, where I would then rewrite the same sort of comments across all of the exercise books. Students might then have looked and observed the comments that they were provided with, but we would then quickly move on to the next unit of work. This did not seem particularly helpful; major misconceptions were being picked up long after the teaching, and I had little evidence that students had actually 'got it' after a single written statement had been provided to them on the strengths and weaknesses of their writing.

"I therefore decided to adopt the practice of 'dot marking', to see if it was sustainable and valuable to judge every student's work, after every lesson, at Key Stage Three. I restructured all of my lessons so that they drove towards a short piece of writing at the end, revolving around a question that I believed would stretch all students, and demand them to explain the main concept of the lesson in their own words. I had previously experimented with 'exit-tickets', where students would write some form of judgement on a Post-it note, but I had found that these were too short. Without dedicating a significant portion of the lesson over to a task, my students were treating this cursorily and I was not left with any sense of the depth of students' understanding. Once I was collecting in students' paragraphs, I dedicated approximately twenty minutes, per class, in a morning to read over all of the paragraphs. Every student was awarded one of three coloured stickers. I regularly mixed the colours up so as to avoid any sense that these were grades. My students have always been in the routine of completing a 'bell task' as they walk into the room. These tasks switched to becoming focused tasks, divided up depending on which colour students had been given. Students were placed into three groups: 'got it', 'almost there' and 'has misunderstood the concept'. The former were offered a question to stretch their thinking further, the 'almost there' students were asked questions of clarification or extension which I would check at some point during the lesson. Those students whose responses I had judged to be weaker were brought together and given a pre-planned explanation to help them overcome their initial misconceptions, and then they were instructed to redraft their paragraphs.

"I found this extremely beneficial. Students' misconceptions, or procedural errors, were being rooted out far earlier, allowing me to treat summative assessment as just that. I put more of my time into making accurate judgements rather than providing formative comments on what should have been summative tasks. I was also making a note in my lesson PowerPoints of the key misconceptions that were arising, which I have then planned ways of avoiding, addressing, or at least discussing when a topic was taught again in a later year.

"The main challenge to providing direction to students in this way was still trying to conform to whole school and departmental expectations of providing written feedback every six lessons. I was saving great sums of time, but on balance, more time was being spent applying additional comments to be seen to be meeting school policies. I worked around this in two ways. At the right juncture, I would provide students with a series of 'WWW/EBI' comments that they could use depending upon the colour their task was given. I followed this up by demonstrating the efficacy of my approach to senior leadership, who have since significantly relaxed their expectations and encouraged a diversification of how feedback is provided to students. I was pleasantly surprised that students did not conflate the colours with grades. This was the case with one or two classes, but when it was explained that this was individualised feedback for specific comments they had made rather than a judgement on their performance, they seemed to be more than satisfied with the approach.

"The one tip that I picked up from this experiment was to plan what I needed to see and provide feedback on, and what I did not. I was finding that there were paragraphs I was reading where I already had a sense or even evidence from the lesson that students had 'got it'. Meanwhile, there were ideas appearing in students' writing that I wished I had sought evidence of in the lesson, and addressed immediately. So, while I have not continued to look at a piece of writing from every student, from every lesson, I have instead planned what I need to provide feedback on for students, and when. This can only make my use of time more efficient, and allow more time to be given over to planning better lessons that address misconceptions I now know to be looking out for."

Checklist

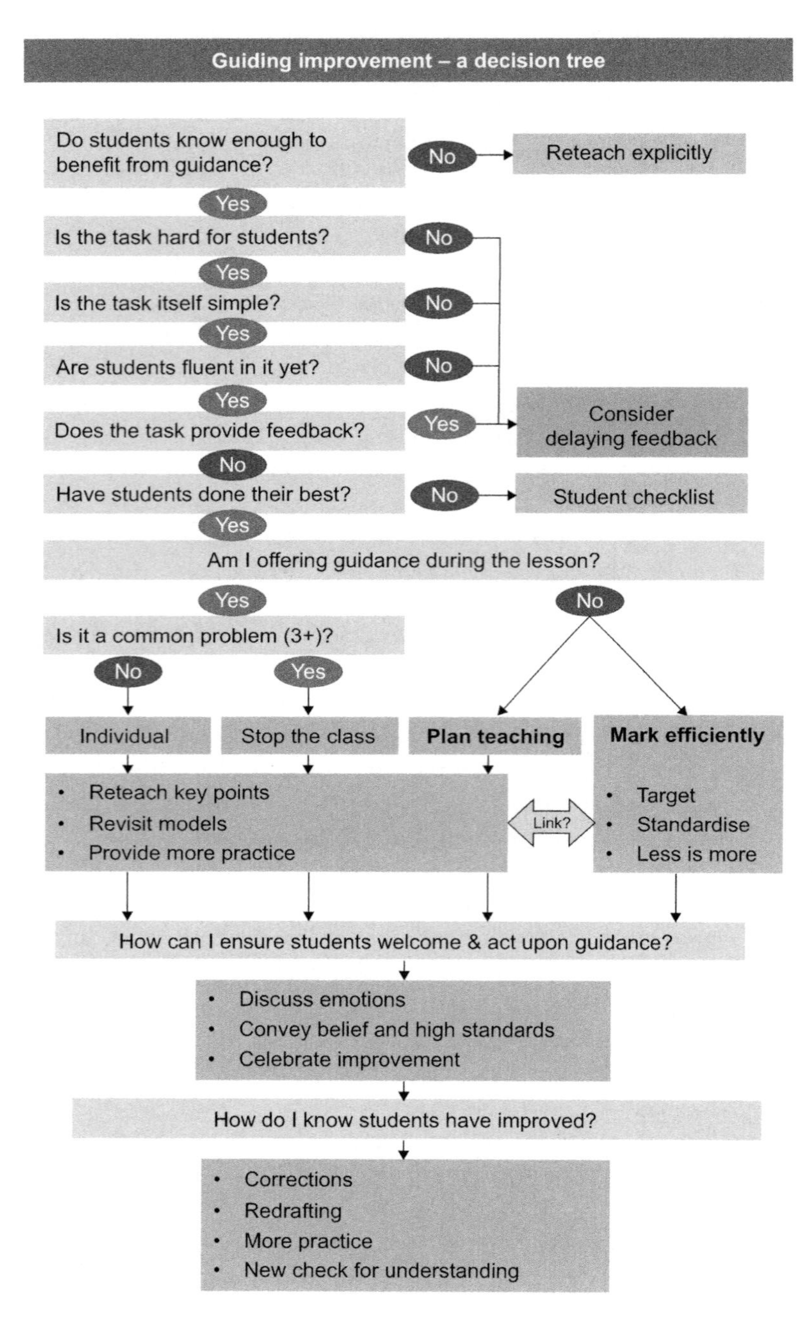

Figure 6.2 Feedback: a decision tree

Checklist: more efficient, effective marking

1 Is the problem further upstream?
- Do students know what success looks like? ☐
- Does the task allow focused feedback? ☐

2 Can students take more responsibility for this work? ☐
3 Can you target your marking on a specific aspect of the work? ☐
4 Can you standardise your approach to marking? ☐
5 Can you limit how much you are writing? ☐
6 Can students improve without individual feedback?
- Give feedback during the lesson ☐
- Reteach ☐
- Revisit model work ☐
- Revise the process ☐
- Give more practice ☐

A great read on this is . . .

Berger, R. (2003). *An ethic of excellence: Building a culture of craftsmanship with students*. Portsmouth, NH: Heinemann.

Ron Berger is a carpenter and an elementary school teacher. This book shows how he creates a culture of craftsmanship and of continual improvement in his classroom.

7 How can we make this work in reality?

The problem

Needs are infinite in schools; resources - of an individual, department or school - are not. The previous six chapters presented evidence, ideas and advice, but as Dylan Wiliam has noted, most teachers are doing all they can to increase student achievement (2011, pp. 18-19). In applying the principles of responsive teaching, two further problems emerge:

1) *When is this to be done?*

Teachers are incredibly busy: days and weeks are full. Even if a change will save time later, initiating it requires time.

2) *How can we create change sustainably and effectively?*

Having set aside time, turning good intentions into lasting habits remains hard. Too often, we try something, like it, struggle to make it sustainable and then leave it behind. To improve students' learning, we need to initiate change and then maintain and refine it.

This chapter suggests ways in which the evidence and principles from previous chapters can be applied by teachers in different roles:

a How can we make time for responsive teaching?
b How can a teacher adopt responsive teaching?
c How can a new teacher adopt responsive teaching?
d What can a mentor do to support responsive teaching?
e What can a head of department do to promote responsive teaching?
f How can professional development support responsive teaching?
g How can senior leaders enable responsive teaching?

A great read on this is . . .

Heath, C. and Heath, D. (2010). *Switch: How to change things when change is hard.* London: Random House.

The Heath brothers turn the evidence into a narrative with three characters: our instincts - the elephant; our rational mind - the rider; and the context we are in - the path. They invite us to plan changes based on a realistic view of how we behave.

a) How can we make time for responsive teaching?

Responsive teaching should reduce workload or allow us to use our time better, but becoming a more responsive teacher demands time to plan and introduce change. This section offers guidance in how to create that time and, more broadly, ways to limit our workload and focus our effort on the most important tasks. The section rests on three beliefs about teaching:

1. There is more that we would like to do for students than there will ever be time for; we cannot complete every desirable task.
2. Not all tasks are equally valuable; completing a worthwhile task may prevent us from doing something even more important.
3. We cannot solve these challenges by creating more time, being more efficient or juggling goals, because the possible tasks will remain unlimited.

This guidance builds on the framework offered by Stephen Covey (2003) to suggest:

1) Be clear what you want to achieve

What are you in school to do? This sounds obvious, but deciding whether your greatest priority is opening students' eyes to the wonders of physics, ensuring every student passes their GCSE or helping your colleagues support students through SATs has important consequences for how you allocate your time.

2) Plan three priorities a week in each role you hold

Set aside a time each week in which to plan your priorities for the following week. Choose three priorities for each role you hold (such as teacher, tutor, mentor, head of department). For example, as a teacher, tutor and literacy coordinator, you would set nine priorities, such as:

- Teacher:
  - Provide feedback on Year 10 essays.
  - Create knowledge organiser for next Year 7 unit.
  - Organise catch-up for absent students in Year 9.
- Form tutor:
  - Support student returning from exclusion.
  - Organise Friday form time activity.
  - Planning meeting for end-of-year trip.
- Literacy coordinator:
  - Finish designing grammar posters.
  - Prepare to present posters in next staff meeting and distribute next week.
  - Audit student spelling in Year 7.

3) Block out time for each goal in the next week

Assign each goal a realistic time over the next week: mark Year 10 essays on Wednesday, finish grammar posters Period 1 on Thursday and so on.

4) Do the tasks in the times allotted (and don't do anything else)

Although prioritising is crucial, the steps above may seem prosaic; the challenge is not in establishing priorities but in sticking to them when buffeted by events. It's easy to write a schedule; it's hard to stick to it when faced with unexpected requests from colleagues and unpredictable demands from students. The most effective response to the pressures of work is not to do more things; it is to stop doing things which seem urgent or are quite important, to allow us to focus on the things which are most important. We need to respond to additional demands on our time in ways which are positive and collegial, but which protect the time designated for our priorities: writing exit tickets, crafting hinge questions, experimenting with feedback. We can:

SEE THE PIE AS FIXED (RATHER THAN ADDING TO IT)

Once a pie is baked, we cannot add more pastry and filling, even if additional guests show up; likewise, once we have a schedule for the week, we won't add extra hours, even if new tasks emerge. The pie is fixed. If new guests arrive, we can either divide the pie differently or send guests away hungry; rather than doing extra hours, we can either reallocate the hours we have or defer or reject those tasks. So, if a new, critical task emerges, it does not become an additional hour's work; we take a less important task and remove, postpone or reduce it. To mix metaphors, as camels, we remove one straw before adding another.

Speaking of the pie as fixed:

- If I submit the SEN return this week, next term's planning will have to slip to next week.
- I'm happy to do this; here's my 'To do' list for this week, what would you suggest I defer?

MAKE THE CONSEQUENCES OF A REQUEST CLEAR

Sometimes, colleagues are unaware of the consequences of their requests. This is particularly common when requests are made to staff across the school; a colleague may not realise the pressures faced by those outside their own department or year group. Making colleagues aware of the steps we will have to take to comply with their request, and the consequences for the rest of our work, may lead them to amend the task or extend the deadline.

Speaking of consequences:

- Filling in this form means going back through every student's data for every unit this year: what's the information you actually need? Is there an easier way to do this?
- Completing this task this week will mean I can't also get all my assessments done as well: which should be the priority?

MAKE A COUNTER-OFFER

If explaining the consequences of a request leaves colleagues unmoved, or we don't have the confidence to give our true answer, we may seek to moderate the task ourselves. We can often provide something which satisfies our colleague without creating additional work: a simpler version or something we've already done.

Speaking of alternatives:

- I'm not sure I'd have time to complete this by Friday, but I'd be happy to discuss it in our team meeting on Tuesday and give you a summary of everyone's thoughts after that.
- Students have their books for revision this week so I can't copy answers for moderation, but I have a range of examples of student work collected across the year already, would you like that?

DON'T DO IT

This is the most powerful approach - although it's one to use judiciously. In the welter of requests colleagues make, not every one matters, and not every one is pursued wholeheartedly. I'm not advocating being deliberately obstructive: if a task really matters, it should be done, whether we like it or not (and we should rearrange our priorities for the week accordingly). Many tasks don't matter, however - or don't matter more than improving our teaching. Gauge colleagues' responses, test the wind, allow the deadline to approach; if colleagues are following the request diligently, do likewise. If not, let it slide and see whether it is resurrected, moderated or forgotten.

Speaking of not doing it:

- How much progress have you made on getting those answers to Steve?
- I was wondering about waiting to complete that - I think it's going to blow over.

5) *Make a habit of staying focused*

This then needs to become a habit, irrespective of the pressure: remaining focused on the things which matter most, remaining politely resistant to those which do not.

A MATTER OF JUDGEMENT

This relies on our judgement as teachers - and the confidence to use it. I'm not advocating evading tasks which are difficult, boring or tedious - some difficult, boring and tedious tasks are vital. I'm advocating making time to improve and sustain ourselves, rather than becoming overwhelmed by less important things. Stephen Covey puts this well:

> You have to decide what your highest priorities are and have the courage - pleasantly, smilingly, nonapologetically - to say 'no' to things. And the way you do that is by having a bigger 'yes' on the inside. The enemy of the 'best' is often the 'good'. Keep in mind that you are always saying 'no' to something. If it isn't to the apparent, urgent things in your life, it is probably to the more fundamental, highly important things.
>
> (2003, pp. 156–157)

ADDRESSING THE UNDERLYING PROBLEMS

The responses described above are individual, but these problems affect all teachers and can be better addressed collectively and systemically. It may be helpful to:

- Check with professional, critical colleagues before beginning a time-consuming task; they may have a better solution.
- Consider how teams can share tasks and support one another in addressing time-consuming challenges.

The discussion in this section is framed for teachers, but it applies equally to middle and senior leaders. Leaders can protect colleagues from the avalanche of possible tasks, help them remain focused on what matters most and encourage them to manage their workload effectively. Preferably, leaders can also keep teachers blissfully ignorant of all the tasks they have deflected (see 7g for more suggestions for senior leaders).

b) How can a teacher adopt responsive teaching?

Teachers develop routines rapidly: how we take the register, distribute sheets and respond to questions becomes a habit. Once created, a habit is hard to change, even if we want to do so (Webb and Sheeran, 2006; Wood and Neal, 2007). In teaching, changing one routine is like knocking down a domino; it affects many other, linked routines. How we ask questions influences how we respond to students' answers, for example. Changing an effective routine is also likely to lead to a dip in performance: a switch in heart surgery techniques at Great Ormond Street Hospital led, initially, to a higher death rate. The change was worthwhile; once surgeons had mastered the technique, the death rate dropped and babies lived longer, fuller lives (Wiliam, 2016), but an initial dip may be unavoidable. Improving as a teacher therefore requires changing our habits and maintaining those changes to get beyond the initial dip and reap the benefits. The process below may help:

1) Identify a need

Pick a class: use the 'Where to start' chart (Figure 0.1, p. 15) and stop at the first issue. Break the problem into something small which can be addressed immediately:

- The unit plan does not specify what students need to know.
- Students' answers do not meet the desired standards.
- Students don't apply feedback.

2) Choose a change

We're busy and change is hard; a big or complicated change is more likely to lead to failure and resistance. Identify a small, manageable change which addresses the problem and fits existing classroom routines:

- Prepare a knowledge organiser for the unit.
- Revisit what a good answer looks like.
- Check student understanding of feedback using a hinge question.

3) Choose a measure

Discussing Problem 4, I mentioned that smart, well-intentioned actions often have unforeseen consequences. We need an objective measure of the impact of our change. Identify something you can measure and record easily which will show the impact on students.

- Can I plan each lesson's key ideas based on the knowledge organiser?
- Do more students use the key features of a good answer?
- What proportion of students answer the hinge question correctly, showing understanding of feedback?

4) Commit to the change

We need to go beyond trying, liking and then ditching changes. We need to allow time to overcome an initial dip: to make the change work and see its merits. Commit to a minimum period - at least a dozen lessons to keep using the change. As Warren Valentine's experience shows (Problem 6), commitment to a technique for some time may teach us something, even if we then adopt a different approach.

- Commit to temporal landmarks such as spending a half-term on the change.
- Commit to students: describe your intentions and rationale to their support and demonstrate your effort.
- Commit to peers: describing your plan creates pressure to pursue it and interest in the results, once they emerge. If you can commit to a common change with several peers, this would be even more powerful, both in changing and in learning from it.

5) Change, refine, evaluate

Set aside extra time to make the change and to evaluate its effect (see Problem 7a). Record the impact each time you try the change. Keep tweaking to make the change fit your routine and your students' needs.

- The knowledge organiser is helpful but needs more detail for a lesson structure: redraft.
- Students identify some answer features but struggle with the writing: model the process, too.
- Students struggle to answer hinge questions about the feedback effectively: give students more specific tasks to help them understand the feedback.

6) Redouble efforts and routinise - or rethink

It's easy to tick something off as 'done' before it has truly become a new routine. If the change is working, set aside more time to improve your approach further. If it's not worked after the initial time you allocated, try a different solution which builds on what you've learned.

- Redouble and routinise: the knowledge organiser was great; write one for every unit next half term.
- Rethink: the problem wasn't student understanding of feedback at all, but their understanding of the expected standard; focus on Problem 3 next half term.

7) Share

Sharing our work provides models for others to learn from; trying to articulate what we've learned pushes our understanding further.

- Write a blog or a piece in your subject association's journal.
- Share with your colleagues at the next department, year team or staff meeting.

8) Repeat

Identify another change to make, following the same process.

c) How can a new teacher adopt responsive teaching?

Skills build on existing skills, so the sequence of learning we follow matters (Ericsson and Pool, 2016). Frances Fuller (1969) argued that new teachers' concerns form a sequence:

1 Should I really be here? Do I belong as a teacher?
2 Will the students really listen? Can I teach?
3 What are students learning?

Fuller argued that new teachers struggle to think about learning if they are concerned about being a teacher or about student behaviour. Recent international research seems to support this: teachers' concerns seem to fit this sequence, and moving along the sequence is linked to better student learning (van de Grift, 2013; van de Grift, Helms-Lorenz and Maulana, 2014). This suggests that responsive teaching is unlikely to be a new teacher's first thought; equally, it suggests that focusing on responsive teaching may be a powerful support for new teachers, offering a way to plan teaching with confidence, understand what students are learning and respond to students' needs, demonstrating that you do belong as a teacher.

I would suggest:

1 Start with Problem 1, clarifying what exactly students are to learn from the unit. Plan units before you plan lessons. Collect suggestions from your colleagues about important representations, explanations and misconceptions and horizon knowledge.
2 Then focus on Problem 2 and identify exactly what you want students to learn; only then can you establish whether it's working.
3 Having addressed these problems, use exit tickets (Problem 4) to identify exactly what students are learning and plan a response.

Whatever you're trying to change:

1 Keep lessons as simple as possible (Problem 2) and create routines. Do challenge students, do try new things, but minimise the number of new things you are trying to do at once to prevent cognitive overload for you and confusion for students. Routines allow you to concentrate on small changes and test their effects: for example, what happens if I use a hinge question to check student understanding (Problem 5) but nothing else changes?
2 Plan changes carefully: write a script for how you will introduce them; rehearse that script. If you are using a new form of feedback for the whole group (Problem 6), prepare your explanation and refine it by practising it aloud, to yourself or a colleague.
3 Find other teachers who are doing responsive teaching well: take specific dilemmas to them and get their advice: 'How would you model solving this problem with students?'
4 Visit as many teachers and schools as you can so you get a sense of 'what success looks like'. Ask them to explain what they are doing, why, and the changes they made in getting where they are today.

5 Record what you're doing and how it's going: forcing yourself to write brief notes somewhere after the lesson is a catalyst for reflection (and so you know how to change it when you return to it a year later).

The guidance for teachers (7b) may help, too.

d) What can a mentor do to support responsive teaching?

Responsive teaching shows teachers how their students are doing. This is particularly important for new teachers: it invites them to take greater responsibility for the effect of their teaching, and to move beyond day-to-day concerns such as 'Will students listen to me?' and 'Have I got enough material to fill the lesson?' One study of experienced mentors suggested that formative assessment was the "most dominant domain of knowledge the mentor needs in order to focus the new teacher on individual student learning" (Athanases and Achinstein, 2003, p. 1501). One experienced mentor they interviewed argued that

> many of my teachers will decide they covered a content/performance standard, and that the majority of students 'got it' and it's time to move on. By working closely to sort and discuss student work, the teacher can make meaningful choices on their next steps/next lessons toward student achievement for each child.
>
> (pp. 1500–1501)

Athanases and Achinstein highlighted that "mentors need command of a wide range of assessment tools and practices in order to help new teachers develop the same" (p. 1501). They advocated presenting evidence to their mentee, prompting reflective conversations about it, listening to the teacher's thinking and moving their attention towards students and their needs:

> While we believe this describes what probably will occur if new teachers are left unguided, we object to the assumption that this must occur. Intervention is possible and potentially effective in interrupting predictable development and in focusing the new teacher's attention early in a career on individual student learning.
>
> (p. 1517)

In a randomised-controlled trial, Jonathan Supovitz (2013) tested the effect of professional learning community meetings focused on linking assessment of student learning with teaching (see 7e). Connecting teachers' choices about teaching to what students learned helped teachers reflect on their teaching, recognise what students understood and led to changes in their teaching and improved student learning. Moreover, teachers' beliefs about what their students can do seem most likely to change based on evidence of students' success (Guskey, 2002).

While a trainee might usefully work through all aspects of this book – from planning to exit tickets, some overall guidance can be offered as well:

1. Begin with the most important section: probably Problem 1, to ensure they have the knowledge to teach; Problem 2, to ensure they have a lesson plan; or Problem 4, to show them what progress looks like.
2. Discuss the evidence in the chapter with your trainee, ensuring they can articulate the underlying principle.
3. Discuss the examples in the book, then offer completion problems; provide partially complete examples from the subject they teach, and work with the trainee to ensure they can apply the principles in a way which suits the subject and the class.

4. Encourage the trainee to practise how they will use their plan: to plan with you, to rehearse explanations, to respond to student misconceptions.
5. Invite trainees to collect evidence of the impact their plan is having, preferably artefacts of student learning.
6. Identify what students are learning (and not learning): try to connect this with how the trainee is succeeding or struggling to put the principle you are working on into practice.
7. Identify a small change which builds on their existing success (see 7f) and gets them closer to applying the principle.
8. Work with them to implement the change.
9. Repeat the process.

e) What can a head of department do to promote responsive teaching?

There's a good case that teaching and learning is best improved by departments. Some things can only be solved at a whole-school level, such as behaviour; others, like lesson planning, can perhaps best be addressed by individual teachers. But it is the department which influences teaching and learning most (Aubrey-Hopkins and James, 2002); it is departments which become the focus for improvement as a school improves (Chapman, 2004). Teachers of different subjects think and interact in different ways (Grossman, Wineburg and Woolworth, 2001; Spillane, 2005); the shared practice of their discipline makes departments distinct "communities of practice" (Harris, 2001; Wenger, 2000, p. 229). Professional learning communities, collegial bodies improving teaching and learning, are usually found in departments (McLaughlin and Talbert, 2001). So how can departments improve teaching and learning?

How: the department as professional learning community

Collective responses to the fundamental challenges facing teachers – What to teach? How best to teach it? – are more powerful. It is in the department where the requisite expertise can be shared (Aubrey-Hopkins and James, 2002). In departments, teachers can reflect on shared experiences; engaged in professional learning communities, they should be able to "increase their professional knowledge and enhance student learning" (Vescio, Ross and Adams, 2007, p. 81). Departments in which students learn more tend to show collegiality, relational trust, teacher learning, shared decision making and a culture of collaboration in which practice is "deprivatised" (Bubb and Earley, 2004; Bryk and Schneider, 2002; Vescio, Ross and Adams, 2007). Perhaps it is thus unsurprising that

> the use of professional learning communities as a means to improve teaching practice and student achievement is a move that educators support and value, as indicated by teachers' perceptions of impact.
>
> (Vescio, Ross and Adams, 2007, p. 88)

Harnessing teachers' collective knowledge and experience should improve student learning in the present and help teachers improve in the longer term, but what should the focus for this collegiality be?

Professional learning towards what?

Collegial communities are only useful if we know what we want (Wiliam, 2007). Departments can work for and against change (Brown, Rutherford and Boyle, 2000; McLaughlin and Talbert, 2001); thus, while a head of department may need to develop collegiality (Harris, 2004), they also need to focus on core goals (Spillane, 2005) and maintain coherence (Sergiovanni, 2005). A review of effective professional learning communities found one feature stood out: "A persistent focus on student learning and achievement by the teachers in the learning communities . . . the collaborative efforts of teachers were focused on meeting the learning needs of their students" (Vescio, Ross and Adams, 2007, p. 87). Conversely: "In the communities where teachers worked together but did not engage in structured work that was highly focused around student learning, similar gains were not evident" (Vescio, Ross and Adams, 2007, p. 87).

A study of one such professional learning community found that its power lay in the use of assessment to connect "the instructional choices that teachers make and the learning outcomes of students." This "helped teachers reflect on their instructional approaches and gain insight into the levels of understanding of their students" and led to changes in their teaching – as identified by external observers – and small, but statistically significant, improvements in student learning (Supovitz, 2013, p. 21). Another study contrasted teachers meeting to discuss teaching and meeting to discuss student work. Discussing teaching, teachers spoke at length, but this led to little critical discussion or insight, whereas discussing student work, teachers

> were constantly monitoring the extent to which there were connections between students' overarching, long-term learning goals, the materials used to assess these goals, and students' related performance.
>
> (Popp and Goldman, 2016, pp. 351–352)

Teachers can collaborate in departments most effectively when they pursue shared goals focused on student learning.

Putting this into practice

How best to achieve this depends on a department's staff, its existing resources (particularly curriculum) and the time available. Time will always be scarce, but a head of department may ask colleagues to sacrifice individual planning time for collaborative planning since it should result not only in the completion of their planning, but in better plans, shared knowledge and resources and professional collegiality. Three approaches may help achieve this:

1) COLLABORATING OVER WHAT TO TEACH

Fundamental questions can be addressed more productively by subject teams. Individual teachers have a range of subject knowledge for teaching and of experiences; the challenge is in creating a structure in which to share this productively. A department can follow the process set out for Problem 1, creating a collective resource for a unit by debating and agreeing:

- The critical knowledge to learn.
- Common student misconceptions.
- Useful images, sources and representations to convey key points.
- Connections to other topics.
- Effective ways to sequence and revisit learning.

This allows every teacher to share their knowledge and experience but avoids creating a strait-jacket, since teachers can use the resulting resource flexibly to suit them and their classes.

2) COLLABORATING OVER HOW TO IMPROVE TEACHING

The aim is not to force teachers to teach identically, but to catalyse reflection about the strengths and weaknesses of teachers' current practice and critical examination of alternatives. Achieving this first requires some kind of common measure: if every student in the year group has completed a similar task, much of the difference in their answers must rest in what (and how) they were taught. Second, teachers need time for common reflection, examining how students responded and trying to identify what it was they did that caused this. How did they explain the topic? How did they allocate time differently? What metaphors did they use when students became stuck? By beginning with a question ('Why did some students answer this question well, others poorly?'), teachers can "move beyond merely sharing what happened in lessons to critical reflection on the teaching-learning process" (Popp and Goldman, 2016, p. 356). Teachers' ability to contrast their approaches with those of their colleagues should allow them to reflect more carefully and more productively, leaving them open to adopt productive ideas willingly. The obvious thing to compare would seem to be exam reviews, but this is unlikely to be particularly helpful, because a summative assessment (or a mock) reveals little about where the gaps in students' knowledge lie (see Introduction): a student may struggle with a specific question for many reasons. Instead, I'd suggest either:

Collective review of an agreed piece of student work Teachers can design half a dozen common exit tickets for a unit (Problem 4), or any agreed piece of student work encapsulating the key ideas. Collective analysis of exit tickets should lead to fruitful discussions, focusing on what students learned and what teachers did differently that may have caused this. Question prompts might include:

- Where did most students struggle?
- What did most students manage well?
- How do student answers differ between classes?

This helps teachers focus upon "the substance represented by the data" and hence "reflect on . . . instructional approaches and gain insight into the levels of understanding of their students" (Supovitz, 2013, p. 21).

CREATING, USING AND ANALYSING MULTIPLE-CHOICE QUESTIONS

Collectively developed multiple choice questions have a range of functions. Designing good hinge questions (Problem 5) is time-consuming and relies on good knowledge of student misconceptions and how students might interpret the questions. Collectively designing half a dozen multiple-choice questions spreads the work and allows teachers to share their knowledge of misconceptions. These questions can be used as hinge questions in the lesson, or as exit tickets, or at any other stage in the learning process (Millar and Hames (2003) show a range of ways teachers have used multiple-choice questions effectively). Subsequent reflection could lead to similar discussions to those conducted with exit tickets, and could also allow the revision and extension of the questions, creating a growing resource.

f) How can professional development support responsive teaching?

> Leaders who are serious about improving the outcomes for students in their schools have to develop the use of formative assessment, both retrospectively, as a way of ensuring that students do not fall behind, and also prospectively, as a way of increasing the pedagogical skills of teachers in the school.
>
> (Wiliam, 2016, p. 126)

Wiliam argues that formative assessment becomes the "vehicle for *delivering*" (p. 133) the range of priorities leaders must pursue. Responsive teaching improves teaching and learning and contributes to metacognition, motivation or applying cognitive science in school. How can whole-school professional development support teachers in this?

What does the evidence suggest?

Effective professional development means focusing "on *what* we want teachers to change, or change about what they do, and we have to understand *how* to support teachers in making these changes" (Wiliam, 2007, p. 188). If Problems 1–6 establish what is to be changed, how can we help teachers to do so? A range of features of effective professional development have been suggested, such as duration, external expertise and subject knowledge (see, for example, Cordingley et al., 2015; Desimone, 2009; Timperley, 2008). The evidence for these features is limited (Wayne et al., 2008), however, particularly because they are adopted by almost all professional development approaches (see, for example, Darling-Hammond, Hyler and Gardner, 2017). Programmes which apply many of these features often struggle to show results (for a summary, see NCEE, 2016; also Jacob, Hill and Corey, 2017). Moreover, they reverse the order suggested by Wiliam (2007): they focus on the how, not the what. It may instead be helpful to focus on three linked approaches to professional development:

1) MEET INDIVIDUALS' NEEDS FOR CHANGE

We need to define *what* our professional development should promote: not just 'responsive teaching', but the specific support our teachers need to teach more responsively. Harland and Kinder (1997) suggested that professional development might contribute to any combination of nine prerequisites for change, in three levels of increasing importance:

1 Resources, information and awareness.
2 Motivation, emotional engagement and institutional support.
3 Knowledge, skills and a congruence of values (between teachers and the proposed change).

To adopt a new practice, it is likely that teachers will need all of these prerequisites, but the effect of a professional development session or programme will vary between teachers and schools, depending on what they already have: for example, a teacher who is motivated, supported and aware of the evidence around responsive teaching may need only a little knowledge or skill to apply it. A session offering the knowledge and skill this teacher needs may provide information and awareness to another teacher without allowing for the values congruence and feelings which would lead to change. The important thing to consider is not, therefore, whether professional development is sustained or led by experts, but what the needs of the individual and the school are, and how professional development could be designed to support them.

2) OFFER INSIGHTS OR STRATEGIES

Professional development approaches which prescribe what to do or provide a body of knowledge seem not to affect teacher practice; those which provide strategies (exemplifying a range of approaches to a problem and supporting teachers to adopt them) or insights ('aha' moments which change teachers' thinking) seem to be much more effective (Kennedy, 2016). The two can be combined: acting on an 'aha' moment may demand new strategies; a changed approach to teaching may lead to new insights.

3) USE BEHAVIOURAL PSYCHOLOGY

> Bart Millar . . . is frustrated by a few of his students, like Robby and Kent, who frequently arrive late and then sit at the back of the room, talking to each other and laughing and disrupting the class.
>
> (Heath and Heath, 2010, p. 186)

What would you do? My first thought wasn't this:

> He bought a used couch and put it right at the front of the classroom. It was immediately obvious that this couch was the cool place to sit – students could slouch and relax instead of sitting at a dorky desk. Suddenly Robby and Kent started getting to class *early* every day so they could 'get a good seat.' *They were volunteering to sit at the front of the classroom*. Genius (p. 187).
>
> (Heath and Heath, 2010, p. 187)

I wouldn't necessarily support this as a behaviour management strategy, but it exemplifies how behavioural psychology encourages us to look at change: focusing on how people really think and behave, not just how we might like them too. The EAST framework offered by the Behavioural Insights Team (2015) is a helpful way to approach this. When asking people to change, make it:

- **Easy:** ask for small, simple changes.
- **Attractive:** make the change desirable and attractive.
- **Social:** peer influence is powerful.
- **Timely:** ask people to change at the right time.

Practice

Any whole-school professional development programme should therefore:

- Meet the needs of the group and individuals.
- Offer insights, strategies or both.
- Encourage change using behavioural psychology.

LEADING TRAINING

The structure of each chapter in the book is like that of a training session presenting a responsive teaching technique:

1. **Frame the problem:** connect the session with teachers' concerns and experiences.
2. **Offer supporting evidence:** justify the strategy you are introducing and explain the rationale, using illustrative studies.
3. **Name the principle.**
4. **Show the principle in practice:** offer examples from different subjects. (This is even more convincing if the examples are from teachers in the school and, better still, if those teachers can lead this section of the session.)
5. **Put the principle into practice:** give teachers time to apply what they have learned, designing or refining their approach for forthcoming lessons.

The goal is to offer both the principle (which teachers will apply to their subject, class and teaching style) and examples (which bring the principle to life as well as offering practical ideas).

This approach meets many of the criteria set out above for effective professional development and behavioural change. For example:

- **Easy:** it helps teachers to apply the ideas introduced.
- **Attractive:** it presents evidence for their value and links them to problems teachers care about.
- **Social:** it offers role models from the school showing the changes are possible.

Sessions like this are only a partial solution, however. Change requires sustained exposure to key ideas and multiple chances to put them into practice to form a habit; sessions like this cannot provide this support. While there is scope for collegial, collaborative learning within the structure described above, the accent is on the expertise of the presenter. Nor does a single session build habits, and it is a change of habit which is needed to achieve a sustained change in teacher practice (see Problem 7b). While the structure suggested above may be helpful in introducing teachers to key ideas, this is just the beginning. Two structures may support ongoing change:

Teacher learning communities Dylan Wiliam (2007) has consistently advocated teacher learning communities as the mechanism for promoting change in teacher practice. He notes the value of gradual change, flexibility for those changing, choice, accountability and support. A typical teacher learning community structure, he and Siobhan Leahy (2014) suggest, might run:

1. **Introduction:** agenda and goals.
2. **Starter:** an introductory activity to focus teachers on their learning.
3. **Feedback:** each teacher reports on the change they committed to in the previous meeting.
4. **New learning:** reading, examining models or discussion of a technique.

5 **Personal action planning:** teachers record a goal for the coming month, to implement or refine a change.
6 **Summary:** have the goals been met?

This resembles the structure suggested above for leading training, but with two critical differences. Action planning asks teachers to commit to specific changes; presenting the results of the action planning and receiving feedback encourages teachers to do what they have committed to, while offering fresh examples with which to understand the principles better.

Functionally, teacher learning communities are simple. A school may adopt them, a head of department allocate time to them, or a teacher establish one with a group of like-minded peers. Wiliam suggests meeting monthly, providing time for teachers to try a new technique between meetings; if teachers have the time needed, they might meet (and therefore act) more quickly. Focusing on one principle for at least a term seems helpful, allowing teachers to understand how it applies and refine their approach to it. Since teachers want to feel they are constantly learning something new, however, it's worth approaching the same ideas in a variety of ways, using a range of representations: examples, texts and individuals' approaches. This allows 'old' content to remain fresh and increases the likelihood teachers will see it in a new light and so refine their approaches.

Teacher learning communities accord well with the EAST framework of behavioural change:

- **Easy:** teachers are asked to try just one change between meetings.
- **Attractive:** the benefits of changes are shown through the way they are presented and the experience of peers in attempting them.
- **Social:** meeting as a group and reporting how changes went challenges teachers to make good on their commitments.
- **Timely:** Wiliam argues that a month is just long enough for teachers to experiment, but not so long that teachers lose focus.

Instructional coaching Instructional coaching, or leverage coaching, is a more intense approach to supporting teachers, developed by Uncommon Schools (Bambrick-Santoyo, 2016). This model provides rapid feedback and support to implement changes. Coaches observe teachers once a week, dropping into lessons unscheduled for around twenty minutes. The same week, they spend twenty minutes offering feedback and practising changes to planning or teaching. An experiment testing a similar model suggests it can cause rapid and significant improvement in teaching (Kraft and Blazar, 2016).

OBSERVATION

While observing, the coach seeks to identify the single change which offers the greatest 'leverage': which has the most potential to improve teaching. For example, a clearer learning objective has greater leverage than a better exit ticket; a good exit ticket relies on a clear objective. (The order of principles in this book forms a logical hierarchy of changes.) Coaches narrow the overall change desired to an action step, a change small enough to be achieved in a week. The observer plans the feedback meeting accordingly.

FEEDBACK

1) Praise The observer focuses on a single, specific episode (ideally, the teacher's success in introducing the last action step) and invites the teacher to reflect on their success:

> We agreed you would ask students to identify a strength and a weakness in each model answer they examined. It was good to see you using that; what has the impact been?

2) Probe The coach seeks to elicit the teacher's thinking to identify the rationale for the next action step. They first ask about the purpose of that phase of the lesson, then identify how closely what happened in the lesson met that purpose. The coach or the teacher may suggest changes which would have better achieved that purpose. The coach might prepare the following questions and suggestions:

> What was the aim of the discussion about different proofs?
> How helpful did it prove in meeting that aim?
> What were the barriers to obtaining useful evidence of student understanding?
> I could see that you aimed to get a sense of what students had understood, but I wasn't sure the evidence was proving useful.
> I wondered whether students needed more thinking time before answering, possibly with a structure for their answer to think about.

3) Action step The coach and teacher agree the action step.

> For challenging questions, give students at least fifteen seconds thinking time.

4) Planning and practice The coach practises the action step with the teacher, to prepare them to act upon it. For a change in planning, the coach and teacher may plan a lesson together; for a change in classroom practice, the teacher will practise it. The coach might say:

> Could you think of a question you'll ask in an upcoming lesson . . . and when you're ready, I'd like you to stand, ask the question, pause, and then nominate someone to answer.

The coach can help the teacher to become comfortable with the action step, and to practise likely challenges.

> OK, this time I'm going to be a very impatient student, because I already have the answer . . .

5) Review The coach and teacher briefly review what has been discussed and promise to visit again the following week to look for the action step.

NEXT OBSERVATION (THE NEXT WEEK): STEP FORWARD OR LOOP BACK

In the next observation, either:

1. The observer sees the suggested change and identifies a new action step.
2. The observer doesn't see the suggested change; they review the action step, establish why it didn't happen and find ways to further support the teacher to implement it.
3. The observer doesn't see the action step taken, because there was no opportunity to use it (for example, if the action step related to discussion and the next observation occurs during a practical lesson).

Leverage observations fit the EAST framework of behavioural change:

- **Easy:** teachers are asked to try a small, manageable change; they are supported in making it happen during the feedback meeting.
- **Attractive:** the discussion during the 'Probe' highlights the need which the action step meets; the observations are low-stakes and relate solely to identifying single improvements.
- **Social:** the commitment is made to the observer, who will return.
- **Timely:** feedback occurs within a day of the observation; having the next observation within a week provides teachers with an incentive to act rapidly.

g) How can senior leaders enable responsive teaching?

Senior leaders can train mentors, guide heads of department and coach individual teachers; they might wish to use any of the guidance sections in this chapter to develop responsive teaching in the school. Most importantly however, senior leaders can create the conditions in the school which make responsive teaching easier and, indeed, natural.

Senior leaders and school culture influence student results significantly. A meta-analysis comparing transformational leadership - inspiring and motivating teachers - and instructional leadership - creating conditions for effective teaching - found the latter to be almost four times more effective in increasing student learning (Robinson, Lloyd and Roqe, 2008). Instructional leaders increased learning when they established goals and expectations, provided strategic resources, planned, coordinated and evaluated teaching, promoted professional development and established a calm and orderly environment; of these priorities, teacher development had the greatest impact. An analysis of how effective school principals in the USA spent their time found that coaching teachers, evaluating them and developing the 'educational programme' led to the greatest benefits (Grissom, Loeb and Master, 2013). Finally, a study of school context - how collegial and well-led a school is - showed it had a powerful impact on the rate at which teachers improve (Kraft and Papay, 2014). Change in teaching relies on the alignment of a number of influences on teaching (Cambridge Assessment, 2017); senior leaders have control over several of them, including who they recruit, how they assess and professional development. School leaders have an important role to play; they can create the conditions in which teachers can pursue responsive teaching in a number of ways:

Create time

Help teachers, mentors and departments protect time to focus on improving teaching. This includes:

- Designating and safeguarding time for professional development sessions and for teams and departments to plan, observe and evaluate together.
- Permit, encourage and challenge teachers to make time to improve by explicitly deprioritising lower-priority tasks and encouraging teachers to stop doing them.
- Model this prioritisation: making time to read, visit classrooms and discuss teaching and learning.

Creating time also means licensing teachers to use that time and providing the support they need to do so. For example, if teachers are to mark less, senior leaders may need to explain to parents the choices teachers are being encouraged to make, and the reasons for them (Wiliam, 2017).

Creating coherence . . .

Many school policies stand in the way of effective responsive teaching. For example:

- Frequent summative assessment reduces the time available for formative assessment; teachers and students focus on performing rather than learning.

- Sharing grades with students emphasises performance, not improvement.
- Specified formats for sharing learning objectives achieves little.
- Marking policies may obstruct effective feedback.

The relevant chapters above may guide revision of school policies, but there is an underlying question of ensuring the school's coherence around a handful of key ideas:

- Making responsive teaching a priority in the school development plan and in teachers' professional development targets.
- Creating a common understanding and language around key features of responsive teaching.
- Aligning policies to this common understanding and language: ensuring that approaches to marking, assessment and teaching and learning complement one another.
- Ensuring that teachers have the resources which will support them (well-designed textbooks and banks of formative assessment questions for example).
- Limiting the role of summative assessment, since it can monopolise the attention and time of students, teachers and parents (see Introduction).

The goal is that what teachers need to do to teach responsively is aligned to what the school expects of their marking, planning and professional development; without alignment, few teachers will be able both to meet the school's priorities and improve.

. . . while leaving space for subject and phase-specificity

While shared understanding and coherence are important, responsive teaching only works when it is adapted thoughtfully to the subject. School leaders can support this by:

- Focusing policies on principles: encouraging departments to adapt their practices as flexibly as they like, providing they remain faithful to those principles.
- Emphasising which elements of school policies are designed to be adapted.
- Ensuring mentoring and coaching is provided by teachers in the same subject or phase.

Remove barriers

Creating time and alignment implies removing barriers, but it's worth emphasising. There are many practical and psychological barriers to improving teaching. Pressure, other priorities and distracting school policies can prevent teachers from dedicating time to the work which matters most; part of the leader's role is removing these barriers. One approach might be to ask teachers, perhaps anonymously, what barriers prevent them from improving their teaching. More specifically, teachers could be asked about the barriers to solving particular problems discussed in the book: 'What barriers prevent students from knowing what success looks like in your subject?'

Create alignment

Teachers need to be aligned to the aim of responsive teaching too. At interview, headteachers could recruit for responsive teachers through:

- Questions which focus on teachers' responses to the problems highlighted in this book.
- Asking candidates to reflect on the way they addressed these problems in interview lessons.
- Asking candidates to respond to student learning during interview lessons.

Conclusion

I taught poorly for a long time: sometimes despite my best efforts, sometimes because my efforts were misguided. When I began to grasp formative assessment better, I was smitten: I saw that, if I could put its principles into practice, I could support my students to achieve the success they deserved. Writing this book has helped me to appreciate the enormous potential cognitive science and formative assessment have to improve our teaching. I hope you find the ideas in the book help you to teach more responsively, and help your students to greater success. Please let me know what you make of them, how you use them, and how they can be improved.

Bibliography

American Association for the Advancement of Science (2017). *AAAS science assessment*, 1st June. http://assessment.aaas.org/topics

Anderson, R., Pichert, J. and Shirey, L. (1983). Effects of the reader's schema at different points in time. *Journal of Educational Psychology*, 75(2), pp. 271–279.

Ariely, D. (2013). *The honest truth about dishonesty: How we lie to everyone – especially ourselves*. New York: Harper.

Athanases, S. and Achinstein, B. (2003). Focusing new teachers on individual and low performing students: The centrality of formative assessment in the mentor's repertoire of practice. *Teachers College Record*, 105(8), pp. 1486–1520.

Aubrey-Hopkins, J. and James, C. (2002). Improving practice in subject departments: The experience of secondary school subject leaders in Wales. *School Leadership & Management*, 22(3), pp. 305–320.

Bahník, Š. and Vranka, M. (2017). Growth mindset is not associated with scholastic aptitude in a large sample of university applicants. *Personality and Individual Differences*, 117, pp. 139–143.

Bailin, S., Case, R., Coombs, J. and Daniels, L. (1999). Conceptualizing critical thinking. *Journal of Curriculum Studies*, 31(3), pp. 285–302.

Baird, J., Andrich, D., Hopfenbeck, T. and Stobart, G. (2017). Assessment and learning: Fields apart? *Assessment in Education: Principles, Policy & Practice*, 24(3), pp. 317–350.

Ball, D. (1993). With an eye on the mathematical horizon: Dilemmas of teaching elementary school mathematics. *The Elementary School Journal*, 93(4), pp. 373–397.

Ball, D., Thames, M. and Phelps, G. (2008). Content knowledge for teaching: What makes it special? *Journal of Teacher Education*, 59(5), pp. 389–407.

Bambrick-Santoyo, P. (2016). *Get better faster: A 90-day pan for coaching new teachers*. San Francisco, CA: John Wiley and Sons.

Bandura, A. (1982). Self-efficacy mechanism in human agency. *American Psychologist*, 37(2), pp. 122–147.

Bandura, A. and Schunk, D. (1981). Cultivating competence, self-efficacy, and intrinsic interest through proximal self-motivation. *Journal of Personality and Social Psychology*, 41(3), pp. 586–598.

Bangert-Drowns, R., Kulik, C., Kulik, J. and Morgan, M. (1991). The instructional effect of feedback in test-like events. *Review of Educational Research*, 61(2), pp. 213–238.

Banks, B. (1991). *The KMP way to learn maths: A history of the early development of the Kent Mathematics Project*. Maidstone, UK: Bertram Banks.

Bart, W. M., Post, T., Behr, M. J. and Lesh, R. (1994). A diagnostic analysis of a proportional reasoning test item: An introduction to the properties of a semi-dense item. *Focus on Learning Problems in Mathematics*, 16(3), pp. 1–11.

Behavioural Insights Team (2015). *The Behavioural Insights Team: Update report 2013–2015*.

Behavioural Insights Team (2017). *The Behavioural Insights Team: Update report 2016–17*.

Bennett, R. (2011). Formative assessment: A critical review. *Assessment in Education: Principles, Policy & Practice*, 18(1), pp. 5–25.

Berger, R. (2003). *An ethic of excellence: Building a culture of craftsmanship with students*. Portsmouth, NH: Heinemann.

Black, P. and Wiliam, D. (1998a). Assessment and classroom learning. *Assessment in Education: Principles, Policy and Practice*, 5(1), pp. 7–74.

Black, P. and Wiliam, D. (1998b). *Inside the black box: Raising standards through classroom assessment*. London: GL Assessment.

Bloom, B. S. (1969). Some theoretical issues relating to educational evaluation. In Tyler, R. W. (ed.), *Educational evaluation: New roles, new means: The 68th yearbook of the national society for the study of education (part II)*, Vol. 68(2). Chicago, IL: University of Chicago Press, pp. 26–50.

Bloom, B. S. (1984). The 2 sigma problem: The search for methods of group instruction as effective as one-to-one tutoring. *Educational Researcher*, 13(6), pp. 4–16.

Brinkman, S., Johnson, S., Codde, J., Hart, M., Straton, J., Mittinty, M. and Silburn, S. (2016). Efficacy of infant simulator programmes to prevent teenage pregnancy: A school-based cluster randomised controlled trial in Western Australia. *The Lancet*, 388(10057), pp. 2264–2271.

Brown, M., Rutherford, D. and Boyle, B. (2000). Leadership for school improvement: The role of the head of department in UK secondary schools. *School Effectiveness and School Improvement*, 11(2), pp. 237–258.

Brown, P., Roediger, H. and McDaniel, M. (2014). *Make it stick*. Cambridge, MA: The Belknap Press of Harvard University Press.

Bryk, A. and Schneider, B. (2002). *Trust in schools*. New York: Russell Sage Foundation.

Bubb, S. and Earley, P. (2004). Why is managing change not easy? In *Managing teacher workload: Workload and wellbeing*. London: PCP/Sage.

Butler, R. (1988). Enhancing and undermining intrinsic motivation: The effects of task-involving and ego-involving evaluation on interest and performance. *British Journal of Educational Psychology*, 58(1), pp. 1–14.

Cambridge Assessment (2017). *A Cambridge approach to improving education: Using international insights to manage complexity*.

Casselman, B. and Atwood, C. (2017). Improving general chemistry course performance through online homework-based metacognitive training. *Journal of Chemical Education*, 94(12), pp. 1811–1821. DOI: 10.1021/acs.jchemed.7b00298.

Chapman, C. (2004). Leadership for improvement in urban and challenging contexts. *London Review of Education*, 2(2), pp. 95–108.

Chi, M. T. H. (2008). Three types of conceptual change: Belief revision, mental model transformation, and categorical shift. In Vosniadou, S. (ed.), *Handbook of research on conceptual change*. Hillsdale, NJ: Erlbaum, pp. 61–82.

Chi, M., Glaser, R. and Rees, E. (1982). Expertise in problem solving. In Sternberg, R. (ed.), *Advances in the psychology of human intelligence*. Hillsdale, NJ: Erlbaum, pp. 7–75.

Christodoulou, D. (2017). *Making good progress: The future of Assessment for Learning*. Oxford: Oxford University Press.

Coe, R. (2013). *Improving education: A triumph of hope over experience*. Centre for Evaluation and Monitoring.

Coe, R., Aloisi, C., Higgins, S. and Elliot Major, L. (2014). *What makes great teaching? Review of the underpinning research*. Sutton Trust.

Coffey, J., Hammer, D., Levin, D. and Grant, T. (2011). The missing disciplinary substance of formative assessment. *Journal of Research in Science Teaching*, 48(10), pp. 1109–1136.

Conway, R. (2017). Reflecting on . . . student responses to feedback. *JMS Reflect* [blog], 18th June. https://jmsreflect.blog/2017/06/18/reflecting-on-student-responses-to-feedback/

Cordingley, P., Higgins, S., Greany, T., Buckler, N., Coles-Jordan, D., Crisp, B., Saunders, L. and Coe, R. (2015). *Developing great teaching: Lessons from the international reviews into effective professional development*. London: Teacher Development Trust.

Covey, S. (2003). *The seven habits of highly effective people*. Carlsbad, CA: Hay House.

Cowan, N. (1999). An embedded-processes model of working memory. In Miyake, A. and Shah, P. (eds.), *Models of working memory: Mechanisms of active maintenance and executive control*. New York: Cambridge University Press, pp. 62–101.

Crooks, T. J. (1988). The impact of classroom evaluation practices on students. *Review of Educational Research*, 58(4), pp. 438–481.

Crouch, C. H., Watkins, J., Fagen, A. P. and Mazur, E. (2007). Peer instruction: Engaging students one-on-one, all at once. *Research-Based Reform of University Physics*, 1(1), pp. 40–95.

Csikszentmihalyi, M. (2002). *Flow: The classic work on how to achieve happiness*. London: Rider.

Darling-Hammond, L., Hammerness, K., Grossman, P., Rust, F. and Shulman, L. (2005). The design of teacher education programs. In Darling-Hammond, L. and Bransford, J. (eds.), *Preparing teachers for a changing world: What teachers should learn and be able to do*. San Francisco: Wiley, Ch. 10.

Darling-Hammond, L., Hyler, M. E. and Gardner, M. (2017). *Effective teacher professional development*. Palo Alto, CA: Learning Policy Institute.

Deans for Impact (2015). *The science of learning*. Austin, TX: Deans for Impact.

Deans for Impact (2016). *Practice with purpose: The emerging science of teacher expertise*. Austin, TX: Deans for Impact.

De Bruyckere, P., Kirschner, P. and Hulshof, C. (2015). *Urban myths about learning and education*. London: Academic Press.

Department for Education (2015). *Final report of the Commission on Assessment without levels*.

Desimone, L. (2009). Improving impact studies of teachers' professional development: Toward better conceptualizations and measures. *Education Researcher*, 38(3), pp. 181–199.

Didau, D. (2014). Why AfL might be wrong, and what to do about. *The learning spy* [blog], 12th March. www.learningspy.co.uk/myths/afl-might-wrong/

Didau, D. and Rose, N. (2016). *What every teacher needs to know about . . . psychology*. Woodbridge: John Catt.

D'Mello, S., Lehman, B., Pekrun, R. and Graesser, A. (2014). Confusion can be beneficial for learning. *Learning and Instruction*, 29, pp. 153–170.

Duit, R. (2009). Bibliography: Students' and teachers' conceptions and science education [online]. http://archiv.ipn.uni-kiel.de/stcse/

Elliott, V., Baird, J., Hopfenbeck, T., Ingram, J., Richardson, J., Coleman, R. Thompson, I., Usher, N. and Zantout, M. (2016). *A marked improvement? A review of the evidence on written marking*. London: Education Endowment Fund.

Ericsson, A. and Pool, R. (2016). *Peak: Secrets from the new science of expertise*. London: Bodley Head.

Facer, J. (2016). Giving feedback the Michaela way. *Reading all the books* [blog], 19th March. https://readingallthebooks.com/2016/03/19/giving-feedback-the-michaela-way/

Feynman, R. (1974). *Cargo cult science* [Commencement address]. Caltech. http://calteches.library.caltech.edu/51/2/CargoCult.htm

Fletcher-Wood, H. (2016). *Ticked off: Checklists for students, teachers and school leaders*. Carmarthen: Crown House.

Fryer, R. (2017). *Management and student achievement: Evidence from a randomized field experiment*. NBER Working Paper No. 23437.

Fuchs, L. S. and Fuchs, D. (1986). Effects of systematic formative evaluation: A meta-analysis. *Exceptional Children*, 53(3), pp. 199–208.

Fuller, F. (1969). Concerns of teachers: A developmental conceptualization. *American Educational Research Journal*, 6(2), pp. 207–226.

Gawande, A. (2010). *The checklist manifesto: How to get things right*. London: Profile.

Gentner, D. (1976). The structure and recall of narrative prose. *Journal of Verbal Learning and Verbal Behavior*, 15(4), pp. 411–418.

Gibbons, R. F. (1975). An account of the secondary mathematics individualized learning experiment. *Mathematics in School*, 4(6), pp. 14–16.

Gibson, S., Oliver, L. and Dennison, M. (2015). *Workload challenge: Analysis of teacher consultation responses*. Department for Education.

Gick, M. and Holyoak, K. (1980). Analogical problem solving. *Cognitive Psychology*, 12(3), pp. 306–355.

Gierl, M., Bulut, O., Guo, Q. and Zhang, X. (2017). Developing, analyzing, and using distractors for multiple-choice tests in education: A comprehensive review. *Review of Educational Research*, 87(6), pp. 1082–1116.

Gipps, C. (1994). *Beyond testing: Towards a theory of educational assessment*. Abingdon: Routledge.

Goodrich, J. (2017). Is formative assessment fatally flawed? Confusing learning & performance. *Improving teaching* [blog comment], 26th March. https://improvingteaching.co.uk/2017/03/26/formative-assessment-flawed-confusing-learning-performance/

Gray, E. and Tall, D. (1994). Duality, ambiguity and flexibility: A proceptual view of simple arithmetic. *The Journal for Research in Mathematics Education*, 26(2), pp. 115–141.

Grissom, J., Loeb, S. and Master, B. (2013). Effective instructional time use for school leaders: Longitudinal evidence from observations of principals. *Educational Researcher*, 42(8), pp. 433–444.

Grossman, P., Wineburg, S. and Woolworth, S. (2001). Toward a theory of teacher community. *The Teachers College Record*, 103, pp. 942–1012.

Guskey, T. (2002). Professional development and teacher change. *Teachers and Teaching: Theory and Practice*, 8(3/4), pp. 381–391.

Haladyna, T., Downing, S. and Rodriguez, M. (2002). A review of multiple-choice item-writing guidelines for classroom assessment. *Applied Measurement in Education*, 15(3), pp. 309–333.

Hammond, K. (2014). The knowledge that 'flavours' a claim: Towards building and assessing historical knowledge on three scales. *Teaching History*, 157, pp. 18–24.

Harland, J. and Kinder, K. (1997). Teachers' continuing professional development: Framing a model of outcomes. *Journal of In-Service Education*, 23(1), pp. 71–84.

Harris, A. (2001). Department improvement and school improvement: A missing link? *British Educational Research Journal*, 27(4), pp. 477–486.

Harris, A. (2004). Distributed leadership and school improvement: Leading or misleading? *Educational Management Administration & Leadership*, 32, pp. 11–26.

Hattie, J. and Timperley, H. (2007). The power of feedback. *Review of Educational Research*, 77(1), pp. 81–112.

Heath, C. and Heath, D. (2010). *Switch: How to change things when change is hard*. London: Random House.

Howe, C. and Abedin, M. (2013). Classroom dialogue: A systematic review across four decades of research. *Cambridge Journal of Education*, 43(3), pp. 325–356.

Hunter, M. C. (1982). *Mastery teaching*. El Segundo, CA: Tip Publications.

Independent Teacher Workload Review Group (2016). *Eliminating unnecessary workload around marking: Report of the Independent Teacher Workload Review Group*. Department for Education.

Jacob, R., Hill, H. and Corey, D. (2017). The impact of a professional development program on teachers' mathematical knowledge for teaching, instruction, and student achievement. *Journal of Research on Educational Effectiveness*, 10(2), pp. 379–407. DOI: 10.1080/19345747.2016.1273411.

Kahneman, D. (2011). *Thinking, fast and slow*. London: Penguin.

Kahneman, D. and Klein, G. (2009). Conditions for intuitive expertise: A failure to disagree. *American Psychologist*, 64(6), pp. 515–526.

Kalyuga, S. and Sweller, J. (2004). Measuring knowledge to optimize cognitive load factors during instruction. *Journal of Educational Psychology*, 96(3), pp. 558–568.

Kennedy, M. (2016). How does professional development improve teaching? *Review of Educational Research*, 86(4), pp. 945–980.

Kingston, N. and Nash, B. (2011). Formative assessment: A meta-analysis and a call for research. *Educational Measurement: Issues and Practice*, 30(4), pp. 28–37.

Kini, T. and Podolsky, A. (2016). *Does teaching experience increase teacher effectiveness? A review of the research*. Palo Alto: Learning Policy Institute.

Kirby, J. (2014). What can we learn from Dylan Wiliam and AfL? *Pragmatic reform* [blog], 30th March. https://pragmaticreform.wordpress.com/2013/03/30/afl/

Kirschner, P. and van Merriënboer, J. (2013). Do learners really know best? Urban legends in education. *Educational Psychologist*, 48(3), pp. 169–183.

Klein, G. (1998). *Sources of power: How people make decisions*. Cambridge, MA: MIT Press.

Kluger, A. and DeNisi, A. (1996). The effects of feedback interventions on performance: A historical review, a meta-analysis, and a preliminary feedback intervention theory. *Psychological Bulletin*, 119(2), pp. 254–284.

Koriat, A. (2007). Metacognition and consciousness. In Zelazo, P., Moscovitch, M. and Thompson, E. (eds.), *The Cambridge handbook of consciousness*. Cambridge: Cambridge University Press, Ch. 11.

Kraft, M. and Blazar, D. (2016). Individualized coaching to improve teacher practice across grades and subjects: New experimental evidence. *Educational Policy*, 31(7), pp. 1033–1068.

Kraft, M. and Papay, J. (2014). Can professional environments in schools promote teacher development? Explaining heterogeneity in returns to teaching experience. *Educational Evaluation and Policy Analysis*, 36(4), pp. 476–500.

Kruger, J. and Dunning, D. (1999). Unskilled and unaware of it: How difficulties in recognizing one's own incompetence lead to inflated self-assessments. *Journal of Personality and Social Psychology*, 77(6), pp. 1121–1134.

Kyun, S., Kalyuga, S. and Sweller, J. (2013). The effect of worked examples when learning to write essays in English literature. *The Journal of Experimental Education*, 81(3), pp. 385–408.

Larkin, J., McDermott, J., Simon, D. and Simon, H. (1980). Expert and novice performance in solving physics problems. *Science*, 208(4450), pp. 1335–1342.

Lemov, D. (2015). *Teach like a champion 2.0*. San Francisco, CA: Jossey-Bass.

Lemov, D., Driggs, C. and Woolway, E. (2016). *Reading reconsidered: A practical guide to rigorous literacy instruction*. San Francisco, CA: Jossey-Bass.

Lin-Siegler, X., Shaenfield, D. and Elder, A. (2015). Contrasting case instruction can improve self-assessment of writing. *Educational Technology Research and Development*, 63, pp. 1–21.

Lipsky, M. (1980[2010]). *Street-level bureaucracy: Dilemmas of the individual in public services*. New York, NY: Russell Sage Foundation.

Livingston, C. and Borko, H. (1989). Expert-novice differences in teaching: A cognitive analysis and implications for teacher education. *Journal of Teacher Education*, 37, pp. 36–42.

Locke, E. A. and Latham, G. P. (2002). Building a practically useful theory of goal setting and task motivation. *American Psychologist*, 57, pp. 705–717.

Loughran, J., Berry, A. and Mulhall, P. (2012). *Understanding and developing science teachers' pedagogical content knowledge: 2nd edition*. Rotterdam: Sense Publishers.

Lu, J., Quoidbach, J., Gino, F., Chakroff, A., Maddux, W. and Galinsky, A. (2017). The dark side of going abroad: How broad foreign experiences increase immoral behavior. *Journal of Personality and Social Psychology*, 112(1), pp. 1–16.

Marshall, B. and Drummond, M. (2006). How teachers engage with Assessment for Learning: Lessons from the classroom. *Research Papers in Education*, 21(2), pp. 133–149.

Massey, C. (2016). Asking year 12, 'what would Figes do?' Using an academic historian as the gold standard for feedback. *Teaching History*, 164, pp. 29–37.

McInerney, L. (2013). The quick, the weird, and the thorough: How I mark student work. *LauraMcInerney.com* [blog], 4th November. https://lauramcinerney.com/2013/11/04/the-quick-the-weird-and-the-thorough-how-i-mark-student-work/

McLaughlin, M. and Talbert, J. (2001). *Professional communities and the work of high school teaching*. Chicago: University of Chicago Press.

Meyer, J. and Land, R. (2003). Threshold concepts and troublesome knowledge 1: Linkages to ways of thinking and practising. In Rust, C. (ed.), *Improving student learning: ten years on*. Oxford: OCSLD.

Millar, R. (2016). Using assessment to drive the development of teaching-learning sequences. In Lavonen, J., Juuti, K., Lampiselkä, J., Uitto, A. and Hahl, K. (eds.), *Electronic proceedings of the ESERA 2015 conference: Science education research: Engaging learners for a sustainable future*, Part 11 (co-ed. Dolin, J. and Kind, P.). Helsinki, Finland: University of Helsinki, pp. 1631–1642. www.esera.org/publications/esera-conference-proceedings/science-education-research-esera-2015/

Millar, R. and Hames, V. (2003). *Using diagnostic assessment to enhance teaching and learning: A study of the impact of research-informed teaching materials on science teachers' practices*. Evidence-Based Practice in Science Education (EPSE) Research Network.

Mittler, P. (1973). Purposes and principles of assessment. In Mittler, P. (ed.), *Assessment for Learning in the mentally handicapped*. Edinburgh, UK: Churchill Livingstone, pp. 1–16.

Morgan, J. (n.d.). Misconceptions. *Resourceaholic* [blog]. www.resourceaholic.com/p/misconceptions.html

Muller, D. (2011). The key to effective educational science videos. *TED*. https://tedxsydney.com/talk/derek-muller-the-key-to-effective-educational-science-videos/

National Trust (n.d.). *50 things to do before you're 11 ¾*. www.nationaltrust.org.uk/documents/50-things-activity-list.pdf

Natriello, G. (1987). The impact of evaluation processes on students. *Educational Psychologist*, 22(2), pp. 155–175.

NCEE (2016). *Does content-focused teacher professional development work? Findings from three institute of education sciences studies*. NCEE Evaluation Brief. National Centre for Education Evaluation and Regional Assistance.

Nelson, M. and Schunn, C. (2008). The nature of feedback: How different types of peer feedback affect writing performance. *Instructional Science*, 37(4), pp. 375–401.

Nuthall, G. (2007). *The hidden lives of learners*. Wellington, NZ: New Zealand Council for Educational Research.

Parkes, J. and Zimmaro, D. (2016). *Learning and assessing with multiple-choice questions in college classrooms*. New York: Routledge.

Pashler, H., Bain, P., Bottge, B., Graesser, A., Koedinger, K., McDaniel, M. and Metcalfe, J. (2007). *Organizing instruction and study to improve student earning* (NCER 2007–2004). Washington, DC: National Center for Education Research, Institute of Education Sciences, U.S. Department of Education.

Perkins, D. and Salomon, G. (1989). Are cognitive skills context-bound? *Educational Researcher*, 18(1), pp. 16–25.

Pershan, M. (2017). Teaching, in general. *Teaching problems* [blog], 15th November. https://problemproblems.wordpress.com/2017/11/15/teaching-in-general/

Petrosino, A., Turpin-Petrosino, C., Hollis-Peel, M. E. and Lavenberg, J. G. (2013). 'Scared Straight' and other juvenile awareness programs for preventing juvenile delinquency. *Cochrane Database of Systematic Reviews*, DOI: 10.1002/14651858.CD002796.pub2.

Phillips, R. (2001). Making history curious: Using Initial Stimulus Material (ISM) to promote enquiry, thinking and literacy. *Teaching History*, 105, pp. 19–25.

Polanyi, M. (1962). *Personal knowledge: Towards a post-critical philosophy*. Abingdon: Routledge.

Popp, J. and Goldman, S. (2016). Knowledge building in teacher professional learning communities: Focus of meeting matters. *Teaching and Teacher Education*, 59, pp. 347–359.

Posner, G., Strike, K., Hewson, P. and Gertzog, W. (1982). Accommodation of a scientific conception: Toward a theory of conceptual change. *Science Education*, 66(2), pp. 211–227.

Potvin, P., Sauriol, É. and Riopel, M. (2015). Experimental evidence of the superiority of the prevalence model of conceptual change over the classical models and repetition. *Journal of Research in Science Teaching*, 52(8), pp. 1082–1108.

Pryor, J. and Crossouard, B. (2010). Challenging formative assessment: Disciplinary spaces and identities. *Assessment & Evaluation in Higher Education*, 35(3), pp. 265–276.

Quigley, A. (2017). The problem with past exam papers. *The confident teacher* [blog], 8th April. www.theconfidentteacher.com/2017/04/the-problem-with-past-exam-papers/

Recht, D. R. and Leslie, L. (1988). Effect of prior knowledge on good and poor readers' memory of text. *Journal of Educational Psychology*, 80, pp. 16–20.

Renkl, A., Hilbert, T. and Schworm, S. (2009). Example-based learning in heuristic domains: A cognitive load theory account. *Educational Psychology Review*, 21(67), DOI: 10.1007/s10648-008-9093-4.

Ritter, S., Damian, R., Simonton, D., van Baaren, R., Strick, M., Derks, J. and Dijksterhuis, A. (2012). Diversifying experiences enhance cognitive flexibility. *Journal of Experimental Social Psychology*, 48(4), pp. 961–964.

Robinson, V., Lloyd, C. and Roqe, K. (2008). The impact of leadership on student outcomes: An analysis of the differential effects of leadership types. *Educational Administration Quarterly*, 44(5), pp. 635–674.

Rodriguez, M. (2005). Three options are optimal for multiple-choice items: A meta-analysis of 80 years of research. *Educational Measurement: Issues and Practice*, 24(2), pp. 3–13.

Rohrer, D. and Taylor, K. (2007). The shuffling of mathematics problems improves learning. *Instructional Science*, 35, pp. 481–498.

Ross, J. and Bruce, C. (2007). Professional development effects on teacher efficacy: Results of randomized field trial. *The Journal of Educational Research*, 101(1), pp. 50–60.

Royal Society of Chemistry (n.d.). *Learn chemistry* [website], www.rsc.org/Learn-Chemistry

Rust, C., Price, M. and O'Donovan, B. (2003). Improving students' learning by developing their understanding of assessment criteria and processes. *Assessment & Evaluation in Higher Education*, 28, pp. 147–164.

Sadler, D. (1989). Formative assessment and the design of instructional systems. *Instructional Science*, 18(2), pp. 119–144.

Schmidt, H. and Rikers, R. (2007). How expertise develops in medicine: Knowledge encapsulation and illness script formation. *Medical Education*, 41, pp. 1133–1139.

Scriven, M. (1967). The methodology of evaluation. In Tyler, R. W., Gagne, R. M. and Scriven, M. (eds.), *Perspectives of curriculum evaluation*. Chicago, IL: Rand McNally, pp. 39–83.

Sealy, C. (2017). The 3D curriculum that promotes remembering. *Primerytimerydotcom* [blog], 28th October. https://primarytimery.com/2017/10/28/the-3d-curriculum-that-promotes-remembering/

Sergiovanni, T. (2005). *Strengthening the heartbeat: Leading and learning together in schools*. San Francisco: Jossey-Bass.

Sfard, A. (1998). On two metaphors for learning and the dangers of choosing just one. *Educational Researcher*, 27(2), pp. 4–13.

Shulman, L. (1987). Knowledge and teaching: Foundations of the new reform. *Harvard Educational Review*, 57(1), pp. 1–23.

Shute, V. (2008). Focus on formative feedback. *Review of Educational Research*, 78(1), pp. 153–189.

Simon, H. and Chase, W. (1973). Skill in chess. *American Scientist*, 61(4), pp. 394–403.

Smith, E. and Gorard, S. (2005). 'They don't give us our marks': The role of formative feedback in student progress. *Assessment in Education*, 12(1), pp. 21–38.

Smith, M. (2017). Cognitive validity: Can multiple-choice items tap historical thinking processes? *American Educational Research Journal*, 54(5), pp. 1256–1287. DOI: 10.3102/0002831217717949.

Soderstrom, N. and Bjork, R. (2015). Learning versus performance: An integrative review. *Perspectives on Psychological Science*, 10(2), pp. 176–199.

Spillane, J. (2005). Primary school leadership practice: How the subject matters. *School Leadership & Management*, 25(94), pp. 383–397.

Stein, M., Engle, R., Smith, M. and Hughes, E. (2008). Orchestrating productive mathematical discussions: Five practices for helping teachers move beyond show and tell. *Mathematical Thinking and Learning*, 10(4), pp. 313-340.

Stobart, G. and Stoll, L. (2005). The key stage 3 strategy: What kind of reform is this? *Cambridge Journal of Education*, 35(2), pp. 225–238. DOI: 10.1080/03057640500146906

Strachan, S. (2017). Why I love . . . whole class feedback & other time-saving feedback strategies. *Susanenglish* [blog], 23rd October. https://susansenglish.wordpress.com/2017/10/23/why-i-lovewhole-class-feedback-other-time-saving-feedback-strategies/

Strong, M., Gargani, J. and Hacifazlioğlu, O. (2011). Do we know a successful teacher when we see one? Experiments in the identification of effective teachers. *Journal of Teacher Education*, 62(4), pp. 367–382.

Supovitz, J. (2013). *The linking study: An experiment to strengthen teachers' engagement with data on teaching and learning*. CPRE Working Papers.

Swaffield, S. (2009). *The misrepresentation of Assessment for Learning – and the woeful waste of a wonderful opportunity*. AAIA National Conference, Bournemouth.

Sweller, J. (1988). Cognitive load during problem solving: Effects on learning. *Cognitive Science*, 12, pp. 257–285.

Sweller, J., Ayres, P., Kalyuga, S. and Chandler, P. (2003). The expertise reversal effect. *Educational Psychologist*, 38(1), pp. 23-31.

Sweller, J., van Merriënboer, J. and Paas, F. (1998). Cognitive architecture and instructional design. *Educational Psychology Review*, 10, pp. 251–296.

Tharby, A. (2017). Connecting and organising knowledge in English literature. *Reflecting English* [blog], 23rd September. https://reflectingenglish.wordpress.com/2017/09/23/connecting-and-organising-knowledge-in-english-literature/

Timperley, H. (2008). Teacher professional learning and development. *Educational Practices*, 18. International Academy of Education.

Treadaway, M. (2015). Why measuring pupil progress involves more than taking a straight line. *Education datalab* [blog], 5th March. https://educationdatalab.org.uk/2015/03/why-measuring-pupil-progress-involves-more-than-taking-a-straight-line/

van de Grift, W. (2013). Measuring teaching quality in several European countries. *School Effectiveness and School Improvement*, 25(3), pp. 295–311.

van de Grift, W., Helms-Lorenz, M. and Maulana, R. (2014). Teaching skills of student teachers: Calibration of an evaluation instrument and its value in predicting student academic engagement. *Studies in Educational Evaluation*, 43, pp. 150–159.

van Merriënboer, J. and Sweller, J. (2005). Cognitive load theory and complex learning: Recent developments and future directions. *Educational Psychology Review*, 17(2), pp. 147–177.

Vescio, V., Ross, D. and Adams, A. (2007). A review of research on the impact of professional learning communities on teaching practice and student learning. *Teaching and Teacher Education*, 24, pp. 80–91.

Wayne, A., Yoon, K., Zhu, P., Cronen, S. and Garet, M. (2008). Experimenting with teacher professional development: Motives and methods. *Educational Researcher*, 37(8), pp. 469–479.

Webb, T. and Sheeran, P. (2006). Does changing behavioral intentions engender behavior change? A meta-analysis of the experimental evidence. *Psychological Bulletin*, 132(2), pp. 249–268. DOI: 10.1037/0033-2909.132.2.249.

Wenger, E. (2000). Communities of practice and social learning systems. *Organization*, 7, pp. 225–247.

Westerman, D. (1991). Expert and novice teacher decision making. *Journal of Teacher Education*, 42(4), pp. 292–305.

White, B. and Frederiksen, J. (1998). Inquiry, modeling, and metacognition: Making science accessible to all students. *Cognition and Instruction*, 16(1), pp. 3–118.

Whitehead, A. (1911). *An introduction to mathematics*. New York: Henry Holt.

Wiliam, D. (2007). Content then process: Teacher learning communities in the service of formative assessment. In Reeves, D. B. (ed.), *Ahead of the curve: The power of assessment to transform teaching and learning*. Bloomington, IN: Solution Tree, pp. 183-204.

Wiliam, D. (2010). What counts as evidence of educational achievement? The role of constructs in the pursuit of equity in assessment. *Review of Research in Education*, 34, pp. 254–284.

Wiliam, D. (2011). *Embedded formative assessment*. Bloomington, IN: Solution Tree.

Wiliam, D. (2013a). Assessment: The bridge between teaching and learning. Voices from the Middle. National Council of Teachers of English. 21(2). pp. 15–20.

Wiliam, D. (2013b). Formative assessment in mathematics: Opportunities and challenges, presented to Teachers College, Columbia University, New York, October 2013. www.dylanwiliam.org/Dylan_Wiliams_website/Presentations.html

Wiliam, D. (2016). *Leadership for teacher learning: Creating a culture where all teachers improve so that all students succeed*. West Palm Beach, FL: Learning Sciences International.

Wiliam, D. (2017). Assessment, marking and feedback. In Hendrick, C. and McPherson, R. (eds.), *What does this look like in the classroom? Bridging the gap between research and practice*. Woodbridge: John Catt.

Wiliam, D. (2018). *Creating the schools our children need: Why what we're doing right now won't help much and what we can do instead*. Learning Sciences International.

Wiliam, D. and Black, P. (1996). Meanings and consequences: A basis for distinguishing formative and summative functions of assessment? *British Educational Research Journal*, 22(5), pp. 537–548.

Wiliam, D. and Leahy, S. (2014). *Sustaining formative assessment with teacher learning communities*. Learning Sciences, Dylan Wiliam Center.

Willingham, D. (2004). Ask the cognitive scientist: The privileged status of story. *American Educator*, Summer.

Willingham, D. (2009). *Why don't students like school? A cognitive scientist answers questions about how the mind works and what it means for the classroom*. San Francisco, CA: Jossey-Bass.

Wittwer, J. and Renkl, A. (2010). How effective are instructional explanations in example-based learning? A meta-analytic review. *Educational Psychology Review*, 22(4), pp. 393–409.

Wood, W. and Neal, D. (2007). A new look at habits and the habit: Goal interface. *Psychological Review*, 114(4), pp. 843–863.

Wylie, C. and Wiliam, D. (2007). Analyzing diagnostic items: What makes a student response interpretable?, paper presented at the annual meeting of the National Council on Measurement in Education (NCME), Chicago, IL, April 2006. www.dylanwiliam.org/Dylan_Wiliams_website/Papers_files/DIMS%20%28NCME%202007%29.pdf

Yeager, D., Purdie-Vaughns, V., Garcia, J., Apfel, N., Brzustoski, P., Master, A., Hessert, W., Williams, M. and Cohen, G. (2014). Breaking the cycle of mistrust: Wise interventions to provide critical feedback across the racial divide. *Journal of Experimental Psychology: General*, 143(2), pp. 804–824.

Yeager, D. and Walton, G. (2011). Social-psychological interventions in education: They're not magic. *Review of Educational Research*, 81(2), pp. 267–301.

Yin, Y., Shavelson, R., Ayala, C., Ruiz-Primo, M., Brandon, P., Furtak, E., Tomita, M. and Young, D. (2008). On the impact of formative assessment on student motivation, achievement, and conceptual change. *Applied Measurement in Education*, 21(4), pp. 335–359.

Young, M. (2014a). The progressive case for a subject-based curriculum. In Young, M., Lambert, D., Roberts, C. and Roberts, M. (eds.), *Knowledge and the future school: Curriculum and social justice*. London: Bloomsbury, Ch. 4.

Young, M. (2014b). Knowledge, curriculum and the future school. In Young, M., Lambert, D., Roberts, C. and Roberts, M. (eds.), *Knowledge and the future school: Curriculum and social justice*. London: Bloomsbury, Ch. 1.

Zhu, X. and Simon, H. (1987). Learning mathematics from examples and by doing. *Cognitive Instruction*, 4, pp. 137–166.

Zimmerman, B. (2002). Becoming a self-regulated learner: An overview. *Theory into Practice*, 41(2), pp. 64–70.

Index

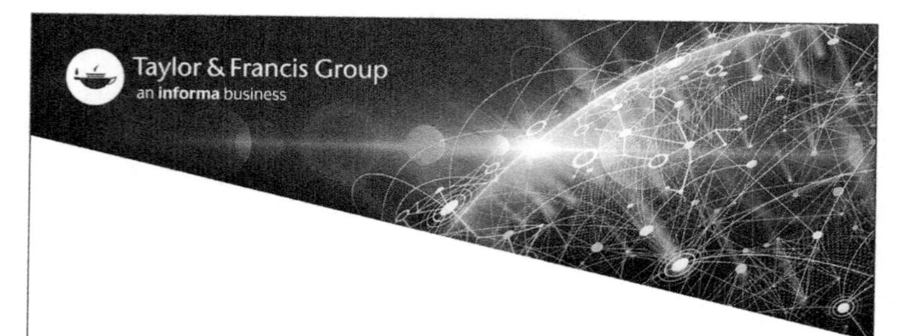

Taylor & Francis eBooks

www.taylorfrancis.com

A single destination for eBooks from Taylor & Francis with increased functionality and an improved user experience to meet the needs of our customers.

90,000+ eBooks of award-winning academic content in Humanities, Social Science, Science, Technology, Engineering, and Medical written by a global network of editors and authors.

TAYLOR & FRANCIS EBOOKS OFFERS:

- A streamlined experience for our library customers
- A single point of discovery for all of our eBook content
- Improved search and discovery of content at both book and chapter level

REQUEST A FREE TRIAL

support@taylorfrancis.com